T0275786

LONDON MATHEMATICAL SOCIETY LECTURE NOTE SERIES

Managing Editor: Professor J.W.S. Cassels, Department of Pure Mathematics and Mathematical Statistics, University of Cambridge, 16 Mill Lane, Cambridge CB2 1SB, England

The books in the series listed below are available from booksellers, or, in case of difficulty, from Cambridge University Press.

34 Representation theory of Lie groups, M.F. ATIYAH *et al*
36 Homological group theory, C.T.C. WALL (ed)
39 Affine sets and affine groups, D.G. NORTHCOTT
40 Introduction to H_p spaces, P.J. KOOSIS
43 Graphs, codes and designs, P.J. CAMERON & J.H. VAN LINT
45 Recursion theory: its generalisations and applications, F.R. DRAKE & S.S. WAINER (eds)
46 p-adic analysis: a short course on recent work, N. KOBLITZ
49 Finite geometries and designs, P. CAMERON, J.W.P. HIRSCHFELD & D.R. HUGHES (eds)
50 Commutator calculus and groups of homotopy classes, H.J. BAUES
51 Synthetic differential geometry, A. KOCK
57 Techniques of geometric topology, R.A. FENN
59 Applicable differential geometry, M. CRAMPIN & F.A.E. PIRANI
62 Economics for mathematicians, J.W.S. CASSELS
66 Several complex variables and complex manifolds II, M.J. FIELD
69 Representation theory, I.M. GELFAND *et al*
74 Symmetric designs: an algebraic approach, E.S. LANDER
76 Spectral theory of linear differential operators and comparison algebras, H.O. CORDES
77 Isolated singular points on complete intersections, E.J.N. LOOIJENGA
78 A primer on Riemann surfaces, A.F. BEARDON
79 Probability, statistics and analysis, J.F.C. KINGMAN & G.E.H. REUTER (eds)
80 Introduction to the representation theory of compact and locally compact groups, A. ROBERT
81 Skew fields, P.K. DRAXL
82 Surveys in combinatorics, E.K. LLOYD (ed)
83 Homogeneous structures on Riemannian manifolds, F. TRICERRI & L. VANHECKE
86 Topological topics, I.M. JAMES (ed)
87 Surveys in set theory, A.R.D. MATHIAS (ed)
88 FPF ring theory, C. FAITH & S. PAGE
89 An F-space sampler, N.J. KALTON, N.T. PECK & J.W. ROBERTS
90 Polytopes and symmetry, S.A. ROBERTSON
91 Classgroups of group rings, M.J. TAYLOR
92 Representation of rings over skew fields, A.H. SCHOFIELD
93 Aspects of topology, I.M. JAMES & E.H. KRONHEIMER (eds)
94 Representations of general linear groups, G.D. JAMES
95 Low-dimensional topology 1982, R.A. FENN (ed)
96 Diophantine equations over function fields, R.C. MASON
97 Varieties of constructive mathematics, D.S. BRIDGES & F. RICHMAN
98 Localization in Noetherian rings, A.V. JATEGAONKAR
99 Methods of differential geometry in algebraic topology, M. KAROUBI & C. LERUSTE
100 Stopping time techniques for analysts and probabilists, L. EGGHE
101 Groups and geometry, ROGER C. LYNDON
103 Surveys in combinatorics 1985, I. ANDERSON (ed)
104 Elliptic structures on 3-manifolds, C.B. THOMAS
105 A local spectral theory for closed operators, I. ERDELYI & WANG SHENGWANG
106 Syzygies, E.G. EVANS & P. GRIFFITH

107 Compactification of Siegel moduli schemes, C-L. CHAI
108 Some topics in graph theory, H.P. YAP
109 Diophantine analysis, J. LOXTON & A. VAN DER POORTEN (eds)
110 An introduction to surreal numbers, H. GONSHOR
111 Analytical and geometric aspects of hyperbolic space, D.B.A.EPSTEIN (ed)
112 Low-dimensional topology and Kleinian groups, D.B.A. EPSTEIN (ed)
113 Lectures on the asymptotic theory of ideals, D. REES
114 Lectures on Bochner-Riesz means, K.M. DAVIS & Y-C. CHANG
115 An introduction to independence for analysts, H.G. DALES & W.H. WOODIN
116 Representations of algebras, P.J. WEBB (ed)
117 Homotopy theory, E. REES & J.D.S. JONES (eds)
118 Skew linear groups, M. SHIRVANI & B. WEHRFRITZ
119 Triangulated categories in the representation theory of finite-dimensional algebras, D. HAPPEL
121 Proceedings of *Groups - St Andrews 1985*, E. ROBERTSON & C. CAMPBELL (eds)
122 Non-classical continuum mechanics, R.J. KNOPS & A.A. LACEY (eds)
124 Lie groupoids and Lie algebroids in differential geometry, K. MACKENZIE
125 Commutator theory for congruence modular varieties, R. FREESE & R. MCKENZIE
126 Van der Corput's method of exponential sums, S.W. GRAHAM & G. KOLESNIK
127 New directions in dynamical systems, T.J. BEDFORD & J.W. SWIFT (eds)
128 Descriptive set theory and the structure of sets of uniqueness, A.S. KECHRIS & A. LOUVEAU
129 The subgroup structure of the finite classical groups, P.B. KLEIDMAN & M.W.LIEBECK
130 Model theory and modules, M. PREST
131 Algebraic, extremal & metric combinatorics, M-M. DEZA, P. FRANKL & I.G. ROSENBERG (eds)
132 Whitehead groups of finite groups, ROBERT OLIVER
133 Linear algebraic monoids, MOHAN S. PUTCHA
134 Number theory and dynamical systems, M. DODSON & J. VICKERS (eds)
135 Operator algebras and applications, 1, D. EVANS & M. TAKESAKI (eds)
136 Operator algebras and applications, 2, D. EVANS & M. TAKESAKI (eds)
137 Analysis at Urbana, I, E. BERKSON, T. PECK, & J. UHL (eds)
138 Analysis at Urbana, II, E. BERKSON, T. PECK, & J. UHL (eds)
139 Advances in homotopy theory, S. SALAMON, B. STEER & W. SUTHERLAND (eds)
140 Geometric aspects of Banach spaces, E.M. PEINADOR and A. RODES (eds)
141 Surveys in combinatorics 1989, J. SIEMONS (ed)
142 The geometry of jet bundles, D.J. SAUNDERS
143 The ergodic theory of discrete groups, PETER J. NICHOLLS
144 Introduction to uniform spaces, I.M. JAMES
145 Homological questions in local algebra, JAN R. STROOKER
146 Cohen-Macaulay modules over Cohen-Macaulay rings, Y. YOSHINO
147 Continuous and discrete modules, S.H. MOHAMED & B.J. MÜLLER
148 Helices and vector bundles, A.N. RUDAKOV et al
149 Solitons, nonlinear evolution equations and inverse scattering, M.A. ABLOWITZ & P.A. CLARKSON
150 Geometry of low-dimensional manifolds 1, S. DONALDSON & C.B. THOMAS (eds)
151 Geometry of low-dimensional manifolds 2, S. DONALDSON & C.B. THOMAS (eds)
152 Oligomorphic permutation groups, P. CAMERON
153 L-functions and arithmetic, J. COATES & M.J. TAYLOR (eds)
154 Number theory and cryptography, J. LOXTON (ed)
155 Classification theories of polarized varieties, TAKAO FUJITA
156 Twistors in mathematics and physics, T.N. BAILEY & R.J. BASTON (eds)
158 Geometry of Banach spaces, P.F.X. MÜLLER & W. SCHACHERMAYER (eds)
159 Groups St Andrews 1989 Volume 1, C.M. CAMPBELL & E.F. ROBERTSON (eds)
160 Groups St Andrews 1989 Volume 2, C.M. CAMPBELL & E.F. ROBERTSON (eds)

London Mathematical Society Lecture Note Series. 160

Groups St Andrews 1989
Volume 2

Edited by
C.M. Campbell
University of St Andrews
and
E.F. Robertson
University of St Andrews

The right of the University of Cambridge to print and sell all manner of books was granted by Henry VIII in 1534. The University has printed and published continuously since 1584.

CAMBRIDGE UNIVERSITY PRESS
Cambridge
New York Port Chester Melbourne Sydney

CAMBRIDGE UNIVERSITY PRESS
Cambridge, New York, Melbourne, Madrid, Cape Town, Singapore, São Paulo

Cambridge University Press
The Edinburgh Building, Cambridge CB2 2RU, UK

Published in the United States of America by Cambridge University Press, New York

www.cambridge.org
Information on this title: www.cambridge.org/9780521406697

© Cambridge University Press 1991

This publication is in copyright. Subject to statutory exception
and to the provisions of relevant collective licensing agreements,
no reproduction of any part may take place without
the written permission of Cambridge University Press.

First published 1991

A catalogue record for this publication is available from the British Library

ISBN-13 978-0-521-40669-7 paperback
ISBN-10 0-521-40669-2 paperback

Transferred to digital printing 2005

CONTENTS

VOLUME II

	Preface	ix
	Introduction	xi
17.	A survey of recent results on projective representations of the symmetric groups J F Humphreys	251
18.	Some applications of graded diagrams in combinatorial group theory A Ivanov & A Yu Ol'shanskii	258
19.	Rational growth of wreath products D L Johnson	309
20.	The modular group and generalized Farey graphs G A Jones, D Singerman & K Wicks	316
21.	On the n-centre of a group L-C Kappe & M L Newell	339
22.	Existentially closed finitary linear groups O H Kegel & D Schmidt	353
23.	Permutability and subnormality of subgroups R R Maier	363
24.	Some applications of powerful p-groups A Mann	370
25.	Combinatorial aspects of finitely generated virtually free groups T Müller	386

26. Problems in loop theory for group theorists
M Niemenmaa 396

27. Sylow theory of CC-groups : a survey
J Otal & J M Pena 400

28. On the minimal number of generators of certain groups
L Ribes & K Wong 408

29. Lie properties of modular group algebras
E Rips & A Shalev 422

30. Observations on a conjecture of Hans Zassenhaus
K W Roggenkamp 427

31. The Fibonacci groups revisited
R M Thomas 445

32. Galois groups
J G Thompson 455

33. Infinite simple permutation groups - a survey
J K Truss 463

34. Polynomial 2-cocyles
L R Vermani 485

CONTENTS OF VOLUME I

	Preface	ix
	Introduction	xi
1.	Triply factorized groups B Amberg	1
2.	An introduction to a class of two relator groups I L Anshel	14
3.	An infinite family of nonabelian simple table algebras not induced by finite nonabelian simple groups Z Arad & H Blau	29
4.	Horace Y Mochizuki: In Memoriam S Bachmuth	38
5.	Bounds on character degrees and class numbers of finite non-abelian simple groups E A Bertram & M Herzog	46
6.	Finite presentability and Heisenberg representations C J B Brookes	52
7.	On nilpotent groups acting on Klein surfaces E Bujalance & G Gromadzki	65
8.	Some algorithms for polycyclic groups F B Cannonito	76
9.	On the regularity conditions for coloured graphs A F Costa	84

10. Multiplet classification of highest weight modules over quantum universal enveloping algebras: the Uq(SL(3,C)) example
V K Dobrev 87

11. Solutions of certain sets of equations over groups
M Edjvet 105

12. Generalizing algebraic properties of Fuchsian groups
B Fine & G Rosenberger 124

13. A theorem on free products of special abelian groups
A M Gaglione & H V Waldinger 148

14. Schur algebras and general linear groups
J A Green 155

15. On Coxeter's groups $G^{p,q,r}$
L C Grove & J M McShane 211

16. Integral dimension subgroups
N D Gupta 214

PREFACE

This is the second of two volumes of the Proceedings of Groups - St Andrews 1989. There is a full contents of both volumes and those papers written by authors with the name of the first author lying in the range H-V. Contained in this part are the papers of three of the main speakers namely O H Kegel, A Yu Ol'shanskii and J G Thompson. Kegel's article is written up jointly with D Schmidt while Ol'shanskii's article is written up jointly with his colleague A Ivanov. We would like especially to thank these main speakers for their contributions both to the conference and to this volume.

INTRODUCTION

An international conference 'Groups - St Andrews 1989' was held in the Mathematical Institute, University of St Andrews, Scotland during the period 29 July to 12 August 1989. A total of 293 people from 37 different countries registered for the conference. The initial planning for the conference began in July 1986 and in the summer of 1987 invitations were given to Professor J A Green (Warwick), Professor N D Gupta (Manitoba), Professor O H Kegel (Freiburg), Professor A Yu Ol'shanskii (Moscow) and Professor J G Thompson (Cambridge). They all accepted our invitation and gave courses at the conference of three or four lectures. We were particularly pleased that Professor Ol'shanskii was able to make his first visit to the West. The above courses formed the main part of the first week of the conference. All the above speakers have contributed articles based on these courses to the Proceedings.

In the second week of the conference there were fourteen one-hour invited survey lectures and a CAYLEY workshop with four main lectures. In addition there was a full programme of research seminars. The remaining articles in the two parts of the Proceedings arise from these invited lectures and research seminars.

The two volumes of the Proceedings of Groups - St Andrews 1989 are similar in style to 'Groups - St Andrews 1981' and 'Proceedings of Groups - St Andrews 1985' both published by Cambridge University Press in the London Mathematical Society Lecture Notes Series. Rather than attempt to divide the two parts by an inevitably imprecise division by subject area we have divided the two parts by author name, the first part consisting of papers with author names beginning A-G and the second part author names beginning H-V.

A feature of these volumes is the number of surveys written by leading researchers, in the wide range of group theory covered. From the papers an extensive list of references may be built up and some of the diversity of group theory appreciated.

The computing aspect of group theory was catered for in several ways. As mentioned above there was a series of lectures on CAYLEY and a CAYLEY workshop; there was also a lecture on GAP. Participants were also given access to the Microlaboratory equipped with MAC's, to the University of St Andrews VAX's and to a SUN 3/260 running a wide range of software including CAYLEY, GAP, SPAS and SOGOS. All these facilities were well used by the conference participants.

Groups - St Andrews 1989 received financial support from the Edinburgh Mathematical Society, the London Mathematical Society and the British Council. We gratefully acknowledge this financial support. We would also like to thank Cambridge University Press and the London Mathematical Society for their help with publishing.

We would like to express our thanks to all our colleagues who helped in the running of the conference and in particular Patricia Heggie, John O'Connor and Trevor Walker. We would also like to thank our wives for their help and forbearance. We would like to thank Shiela Wilson for so willingly undertaking the daunting task of typing the two volumes of the Proceedings, having already typed the Proceedings of the previous two St Andrews conferences.

Our final thanks go to those authors who have contributed articles to these volumes. We have edited these articles to produce some uniformity without, we hope, destroying individual styles. For any inconsistency in, and errors introduced by, our editing we take full responsibility.

Colin M Campbell

Edmund F Robertson

A SURVEY OF RECENT RESULTS ON PROJECTIVE REPRESENTATIONS OF THE SYMMETRIC GROUPS

J F HUMPHREYS

University of Liverpool, Liverpool L69 3BX

The basic results on the projective representations of the symmetric groups were obtained by Schur in 1911, [14]. In recent years, there has been considerable interest in this area. The aim of this article is to outline some of the recent developments. The articles by Stembridge [15] and Józefiak [7] and the forthcoming book [5], provide introductions to the subject.

1. Preliminaries

Definition. A *projective representation* of a group G of degree d over a field K is a map $P : G \to GL(d,K)$ such that

(a) $P(1_G) = I_d$;

and

(b) given x and y in G, there is an element $\alpha(x,y)$ in $K^\times$ (the multiplicative group of K) such that

$$P(x)\, P(y) = \alpha(x,y)\, P(xy).$$

When $\alpha(x,y) = 1$ for all x and y, we say that P is a *linear representation* of G.

Using the fact that the matrices $\{P(g) : g \in G\}$ are invertible, the fact that $P(1_G)$ is the identity matrix gives:

(C1) for all g in G,

$$\alpha(g,1) = 1 = \alpha(1,g).$$

Using invertibility again and also associativity of group composition and of matrix multiplication, (b) gives:

(C2) for all x, y and z in G,

$$\alpha(xy, z)\alpha(x,y) = \alpha(x, yz)\alpha(y,z).$$

A map $\alpha : G \times G \to K^\times$ satisfying (C1) and (C2) is a *2-cocycle.*

There is an equivalence relation on the set of 2-cocycles: α is equivalent to β if and only if there is a map $\delta : G \to K^\times$ such that for all x, y in G

$$\alpha(x,y) = \delta(x)\delta(y)\delta(xy)^{-1}\beta(x,y).$$

The equivalence classes of 2-cocycles form a group with multiplication $[\alpha\beta] = [\alpha][\beta]$ where

$$\alpha\beta(x,y) = \alpha(x,y)\beta(x,y).$$

This group is usually denoted $H^2(G,K^\times)$. When K is algebraically closed of characteristic zero, it is known as the *Schur multiplier* and is often denoted M(G).

Under suitable assumptions of K (for example if K is algebraically closed of characteristic zero), the group G has a *K-representation group.* This is a group H with central subgroup A, isomorphic to $H^2(G, K^\times)$, with H/A isomorphic to G such that every projective representation of G over K can be "lifted" to a linear representation of H. In these circumstances, the projective representations of G may be regarded as linear representations of its representation group.

2. The symmetric groups

For the symmetric group S(n), Schur showed in [14] that the multiplier is cyclic of order 2 when $n \geq 4$. Since complex representation groups always exist, there is a group $\tilde{S}(n)$ of order 2n! which is a "double cover" of S(n). Schur provided an explicit matrix representation of $\tilde{S}(n)$ (the *basic representation*) as follows. Consider the complex matrices

$$A = \begin{pmatrix} 0 & i \\ i & 0 \end{pmatrix}; \quad B = \begin{pmatrix} 0 & 1 \\ -1 & 0 \end{pmatrix}; \quad C = \begin{pmatrix} 1 & 0 \\ 0 & -1 \end{pmatrix}.$$

Then

$$A^2 = B^2 = -I; \quad C^2 = I; \quad AB = -BA; \quad AC = -CA; \quad BC = -CB.$$

Now let m be the integer part of (n-1)/2, and define the following $2^m \times 2^m$ matrices:

$$M_{2m+1} = iC^{\otimes m}$$

and for $1 \leq k \leq m$,

$$M_{2k-1} = C^{\otimes(m-k)} \otimes A \otimes I^{\otimes(k-1)};$$

$$M_{2k} = C^{\otimes(m-k)} \otimes B \otimes I^{\otimes(k-1)}.$$

Thus $M_j^2 = -I$ and $M_jM_k = -M_kM_j$ ($j \neq k$). Finally, for $1 \leq k \leq n-1$, let

$$T_k = -\left(\frac{k-1}{2k}\right)^{1/2} M_{k-1} + \left(\frac{k+1}{2k}\right)^{1/2} M_k.$$

It may be checked that these matrices satisfy the relations

$$T_k^2 = -I \qquad (1 \leq k \leq n-1);$$

$$(T_kT_{k+1})^3 = -I \qquad (1 \leq k \leq n-2);$$

$$T_jT_k = -T_kT_j \qquad (1 \leq j, k \leq n-1, \quad |j-k| > 1),$$

and so generate the required representation P_n of $\tilde{S}(n)$ with degree 2^m. This representation is irreducible and *negative* in the sense that the central involution of $\tilde{S}(n)$ is represented as the negative of the identity matrix. Replacing T_k by $-T_k$ gives another representation P_n' which is equivalent to P_n if and only if n is odd.

Schur [14], showed that the irreducible negative representations of $\tilde{S}(n)$ can be indexed by *strict partitions* of n, that is partitions of n with no repeated parts. This indexing is not, however, bijective. If λ has ℓ non-zero parts, there are two irreducible negative representations corresponding to λ if $n-\ell$ is odd. The two irreducibles in this case are *associated*, in that one is obtained from the other by forming the tensor product with the sign representation. There is a unique irreducible associated with λ when $n-\ell$ is even. The basic irreducible negative representation P_n corresponds to the partition (n).

3. The Q-functions

We next give a combinatorial definition of the Q-functions. Let $\lambda = (\lambda_1, \ldots, \lambda_\ell)$ be a strict partition of n with ℓ non-zero parts. *A shifted Young diagram of shape* λ is a diagram with ℓ rows and λ_i nodes in each row with the first node in row i+1 being under the second node in row i.

In order to define the Q-functions, we let **P** denote the ordered alphabet $\{1' < 1 < 2' < 2 < 3' \ldots\}$. The letters 1′, 2′, 3′, ... are *marked* (and 1, 2, 3, ... *unmarked*). The notation $|a|$ is used for the unmarked version of any element a of **P**.

Definition. A *shifted Young tableau of shape* λ is an assignment of elements of **P** to the nodes of a shifted Young diagram of shape λ such that

(i) the entries are weakly increasing along rows and down columns;

(ii) there is at most one occurrence of any given unmarked letter in any given column,

and

(iii) there is at most one occurrence of any given marked letter in each row.

Example. A shifted Young tableau associated with the partition $\lambda = (7, 5, 3, 2, 1)$ is

$$\begin{array}{ccccccc} 1 & 1 & 2' & 2 & 2 & 4 & 5 \\ & 2' & 2 & 3 & 4 & 6 & \\ & & 3 & 5 & 6 & & \\ & & & 5' & 7' & & \\ & & & & 7 & & \end{array}$$

The *content* of a shifted Young tableau T of shape λ is the sequence $\gamma = (\gamma_1, \gamma_2, \ldots)$ where γ_i is the number of entries of T equal to $|i|$. In the above example $\gamma = (2, 5, 2, 2, 3, 2, 2)$. For any set $\{x_1, x_2, \ldots\}$ of indeterminates, let

$$x^T = x^\gamma = x_1^{\gamma_1} x_2^{\gamma_2} \cdots .$$

Definition. Let λ be a strict partition. Define the Q-function $Q_\lambda = \sum_T x^T$, where the sum is over all shifted Young tableaux of shape λ.

The Q-functions are in fact symmetric functions, as was clear from the original definition of Schur. They play a role for projective representations analogous to that played in the linear representation theory by the Schur functions. The Q-functions can also be regarded as a special case of the Hall-Littlewood polynomials. The details of these alternative descriptions and their equivalence may be found in [5].

4. Recent advances

(a) In 1988, Nazarov [11] announced a method to construct the matrices representing the generators of $\tilde{S}(n)$ for the irreducible negative representations associated with the strict partition λ. His method is to define matrices of the appropriate size, and to check that they do indeed satisfy the relations for the group $\tilde{S}(n)$. The details of this construction may be found in [12].

(b) Recursive formulae to calculate the value of a given irreducible negative character at a specific conjugacy class of $\tilde{S}(n)$ in terms of the values of the characters of $\tilde{S}(k)$, for $k < n$ have been obtained by Morris [9] and Morris-Olsson [10]. This is analogous to the Murnaghan-Nakayama formula for linear characters of $S(n)$.

(c) It is possible to write the product of Q-functions $Q_\lambda Q_\mu$ in the form $\Sigma f^{\nu}_{\lambda\mu} Q_\nu$, where the coefficients can be proved to be non-negative integers. In fact Stembridge [15] has given a combinatorial way to calculate the coefficients $f^{\nu}_{\lambda\mu}$, similar to the Littlewood-Richardson rule for linear representations. The coefficients are somewhat complicated to describe, and the proof of Stembridge's result makes use of results of Worley [16].

(d) There is an algebra associated with the negative representations of the groups $\tilde{S}(n)$, analogous to the well-known graded algebra of Grothendieck groups of linear representations of $S(n)$. This is obtained as follows. For $n \geq 4$, let T^0_n be the Grothendieck group of isomorphism classes of finite-dimensional negative representations of $\tilde{S}(n)$, and let T^1_n be the Grothendieck group of isomorphism classes of finite-dimensional negative representations of $\tilde{A}(n)$. Let T^*_n be the $\mathbf{Z}/2$-graded group $T^0_n \oplus T^1_n$. Let L be the ring $\mathbf{Z}[\lambda]/(\lambda^3 - 2\lambda)$, which is $\mathbf{Z}/2$-graded by letting $L^0 = \mathbf{Z} \oplus \rho\mathbf{Z}$, (where $\rho = \lambda^2 - 1$) and $L^1 = \lambda\mathbf{Z}$. Then, with appropriate definitions of T^*_n for values of $n < 4$, $\mathfrak{I} = \oplus_{n\geq 1} T^*_n$ is a graded L-module. In fact $\mathfrak{I}$ is an algebra (induction product) and coalgebra (restriction) so that $\mathfrak{I}$ is a Hopf algebra. This algebra approach to the subject is investigated in Hoffman and Humphreys, [3] and [4].

(e) The representation theory over modular fields is the subject of investigations by several authors. When p = 2, the modular representations of $\tilde{S}(n)$ coincide with those for S(n). The irreducible negative complex characters then provide a useful class of Brauer characters for S(n) (see Benson [1]).

For odd primes p, the assignment of irreducible complex characters to p-blocks has been given by Humphreys [6] in answer to a conjecture of Morris. Alternative proofs of this result have been given by Cabanes [2] and Olsson [13]. Michler and Olsson [8] have recently shown that the Alperin-McKay conjecture holds in $\tilde{S}(n)$ for odd primes.

References

1. D Benson, Spin modules for symmetric groups, *J. London Math. Soc.* (2) **38** (1988), 250-262.
2. M Cabanes, Local structure of the p-blocks of $\tilde{S}(n)$, *Math. Z.* **198** (1988), 519-543.
3. P N Hoffman & J F Humphreys, Hopf algebras and projective representations of $G \wr S_n$ and $G \wr A_n$, *Canad. J. Math.* **38** (1986), 1380-1458.
4. P N Hoffman & J F Humphreys, Primitives in the Hopf algebra of projective S_n-representations, *J. Pure Appl. Algebra* **46** (1987), 155-164.
5. P N Hoffman & J F Humphreys, *Projective representations of the symmetric groups* (Oxford University Press), to appear.
6. J F Humphreys, Blocks of projective representations of the symmetric groups, *J. London Math. Soc.* (2) **33** (1986), 441-452.
7. T Józefiak, Characters of symmetric representations of symmetric groups, *Exposition. Math.* **7** (1989), 193-237.
8. G O Michler & J B Olsson, The Alperin-McKay conjecture holds in the covering groups of the symmetric and alternating groups, p≠2, to appear.
9. A O Morris, The spin representation of the symmetric group, *Canad. J. Math.* **17** (1965), 543-549.
10. A O Morris & J B Olsson, On p-quotients for spin characters, *J. Algebra* **119** (1988), 51-82.
11. M L Nazarov, An orthogonal basis of irreducible projective representations of the symmetric group, *Functional Anal. Appl.* **22** (1988), 77-78.
12. M L Nazarov, Young's orthogonal form of irreducible projective representations of the symmetric group, to appear.
13. J B Olsson, On the blocks of the symmetric and alternating groups and their covering groups, *J. Algebra*, to appear.

14. I Schur, Über die Darstellung der symmetrischen und der alternierenden Gruppe durch gebrochene lineare Substitutionen, *J. Reine Angew. Math.* **139** (1911), 155-250.

15. J R Stembridge, Shifted tableaux and the projective representations of symmetric groups, *Adv. in Math.* **74** (1989), 87-134.

16. D R Worley, *A theory of shifted Young tableaux* (Ph.D. thesis, M.I.T. 1984).

SOME APPLICATIONS OF GRADED DIAGRAMS IN COMBINATORIAL GROUP THEORY

SERGEI V IVANOV & ALEXANDER YU OL'SHANSKII

Moscow State University, Moscow 119899, USSR

This paper is based on four lectures given by the second author at the conference "Groups - St Andrews 1989", which in turn were based on his book [1] and on results obtained recently by the authors and others at the higher algebra seminar of Moscow State University.

1. Diagrams over groups

1.1 van Kampen's lemma. Let G be a group generated by a set $\{a_1, a_2, ...\}$ of generators and a set of defining relations $\{r_1 = 1, r_2 = 1, ...\}$, where r_1, r_2 ... are words in the group alphabet

$$\mathfrak{A} = \{a_1^{\pm 1}, a_2^{\pm 1}, ...\}.$$

Of course, any group G can be described by means of such a presentation

$$G = < a_1, a_2, ... \mid r_1 = 1, r_2 = 1, ... >.$$

Van Kampen [2] discovered a very simple visual demonstration of the deducibility of consequences of defining relations. For example, the equality

$$a^2b\, ab^{-1} = 1$$

follows from relations

$$a^3 = 1 \text{ and } aba^{-1}b^{-1} = 1$$

and the deducibility is pictured in Fig. 1. We read one of the defining words by going around any region and we read the consequence by going along the boundary of the map (following the inverse edge gives the inverse letter).

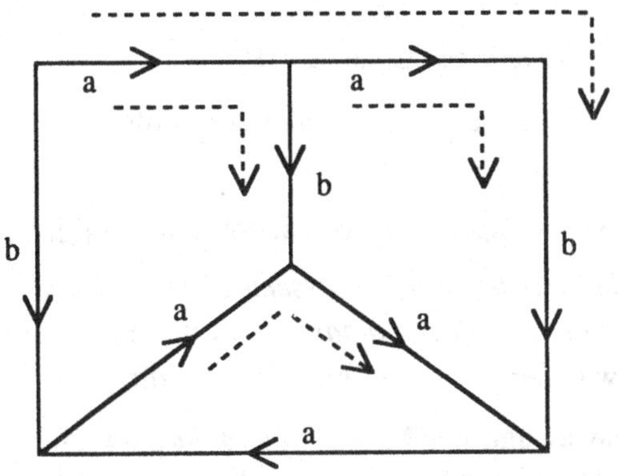

Fig. 1

Let us give some definitions.

A *map* is a finite connected, simply connected planar two-complex. Let

$$\mathfrak{A} = \{a_1^{\pm 1}, a_2^{\pm 1}, \ldots\}$$

be a (group) *alphabet. A diagram* M over $\mathfrak{A}$ is a map which is equipped with a labelling function ϕ from the set of oriented edges (one-cells) of M into $\mathfrak{A}$ with the property that for any edge e of M

$$\phi(e^{-1}) = \phi(e)^{-1}.$$

Let G be a group defined by the presentation

$$G = < \mathfrak{A} \mid \mathfrak{R} >, \qquad (1.1)$$

where $\mathfrak{R}$ is a set of relators of G.

A diagram M over G (more accurately, over the presentation (1.1)) is a diagram M over $\mathfrak{A}$ such that if p is a boundary cycle for a region (= two-cell) of M, then the word $\phi(p)$ is an element of $\mathfrak{R}$.

Of course, the word $\phi(p)$ depends on a choice of an initial vertex on the boundary of the cell and on the choice of direction (clockwise or counterclockwise). However, this causes no trouble, because we can assume that the set of defining words $\mathfrak{R}$ is *symmetrized*:

(a) $r \in \mathfrak{R} \Rightarrow r^{-1} \in \mathfrak{R}$;

(b) $r = uv \in \mathfrak{R} \Rightarrow$ cyclic conjugate $vu \in \mathfrak{R}$;

(c) $r \in \mathfrak{R} \Rightarrow r$ is reduced, i.e. without subwords of the type aa^{-1}, where $a \in \mathfrak{A}$.

The statement of van Kampen's lemma is almost obvious [3].

A relation $w = 1$ *holds in the group* G *presented by* (1.1) *if and only if there exists a diagram* M *over* G *with boundary path* $p = e_1 \dots e_m \in M$ *such that* $\phi(e_1) \dots \phi(e_m) \equiv w$ (where $\equiv$ denotes graphical equality).

Moreover, one may assume that M is a *reduced diagram.* What does this mean? Let us suppose there exists a pair of regions S_1, S_2 with a common edge e such that for boundary paths $p_1 = eq_1$ and $p_2 = eq_2$, beginning with e, the equation $w_1 = w_2$ holds, where the words w_1, w_2 are the labels of p_1 and p_2. Then we can cut out two regions and sew up the hole (see Fig. 2). And so we can assume that the diagram M in van Kampen's lemma does not contain such a pair of contractible regions. In this case, we say that M is a reduced diagram.

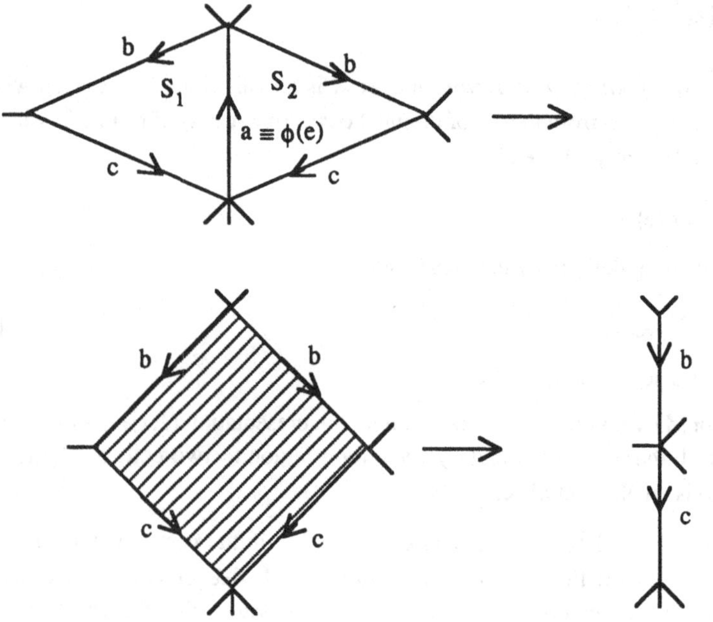

Fig. 2

1.2 Small cancellation conditions. Let $\mathcal{R}$ be a symmetrized set of defining relations of a group G. A common initial segment of two different words of $\mathcal{R}$ is called a *piece*. The most popular condition $C'(\lambda)$ of small cancellation theory says that if u is a piece and $r = uv \in \mathcal{R}$, then $|u| < \lambda|r|$ (where $|w|$ is the length of w).

Example. Let G be the fundamental group of a closed oriented compact Riemann surface of genus 2:

$$G = \langle a_1, a_2, a_3, a_4 \mid a_1 a_2 a_1^{-1} a_2^{-1} a_3 a_4 a_3^{-1} a_4^{-1} = 1 \rangle.$$

Then the symmetrized system $\mathcal{R}$ contains 16 words

$$a_1 a_2 a_1^{-1} a_2^{-1} a_3 a_4 a_3^{-1} a_4^{-1},\ a_2 a_1^{-1} a_2^{-1} a_3 a_4 a_3^{-1} a_4^{-1} a_1,$$

$$\dots$$

$$a_4 a_3 a_4^{-1} a_3^{-1} a_2 a_1 a_2^{-1} a_1^{-1},\ a_1^{-1} a_4 a_3 a_4^{-1} a_3^{-1} a_2 a_1 a_2^{-1}.$$

The subword a_1 is a piece; it is the beginning of words

$$a_1 a_2 a_1^{-1} a_2^{-1} a_3 a_4 a_3^{-1} a_4^{-1}$$

and

$$a_1 a_2^{-1} a_1^{-1} a_4 a_3 a_4^{-1} a_3^{-1} a_2.$$

But $a_1 a_2$ (and any other word u with $|u| = 2$) is not a piece. So the condition $C'(\lambda)$ holds for each $\lambda > 1/8$.

A maximum common subpath of boundary paths of regions in a map will be called an *arc*.

Let S_1 and S_2 be two regions and let p be an arc between them: $p = e_1 \dots e_n$, where $e_1, \dots, e_n$ are edges and $|p| = n$. Suppose that $|p| \geq \lambda|\partial S_1|$, where $|\partial S_1|$ is the length of the perimeter of S_1. Then the labels w_1 and w_2 of the boundary paths of S_1 and S_2 have the common beginning $\phi(p)$, and so $w_1 \equiv w_2$ because of the condition $C'(\lambda)$. Therefore the diagram is not reduced. So the $C'(\lambda)$ condition means that the number of arcs of every interior region of a reduced van Kampen diagram is more than $1/\lambda$.

On the other hand, let us write v, a, r for the numbers of vertices, arcs and regions in a planar map M. (We are not taking into account now the vertices of valency 2.) Then $3v \leq 2a$, because the degree of any vertex is ≥ 3. And Euler's formula gives us the equality $v = a-r+1$. We can eliminate v and obtain the inequality $r > a/3$. Any arc belongs to at most two cells and hence the mean number of arcs for a cell is less than 6. However, the number of arcs of an interior region is more than λ^{-1}, so that if λ is small (say $\lambda^{-1} >> 6$), the number of interior regions is less than the number of exterior regions. Of course, the above arguments are very rough. In fact, it is sufficient to take $\lambda = 1/6$ in small cancellation theory [3]. Anyway, there are many exterior edges in the reduced diagram M if λ is small, and M "looks like" a "tree" (Fig. 3).

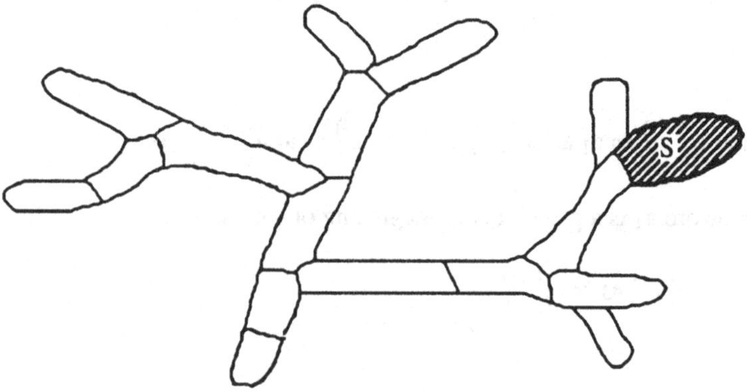

Fig. 3

In particular, there exists a cell S with a "long" external arc p (for instance, $|p| > |\partial S|/2$).

It is clear that we have a basis for *Dehn's algorithm*: if a non-trivial reduced word w equals 1 in G, then w contains a subword u such that u is a subword of a relator $r \in \mathfrak{R}$ and $|u| > 1/2\ |r|$. Thus $r \equiv uv$, where $|u| > |v|$, and we can shorten

$$w \equiv w_1 u w_2$$

to

$$w_1 v^{-1} w_2.$$

In this way, we can reduce the word w to 1. Max Dehn applied the algorithm to the fundamental groups of closed compact Riemann surfaces of genus $g > 1$.

Many other typical results of small cancellation theory are exhibited in the well-known book of R Lyndon and P Schupp [3]. As a rule, small cancellation groups contain non-cyclic free subgroups and plenty of normal subgroups. These and other properties have been extended to the broader and more interesting class of hyperbolic groups. One can find the theory of hyperbolic groups in the important paper of M Gromov [4]. However, the $C'(\lambda)$-condition is a rather simple introduction to the new conditions which will be considered below.

Annular diagrams were introduced by P Schupp for deducibility of conjugacy in groups [5]. Diagrams on other finite 2-complexes interpret homomorphisms of groups [6].

2. Periodic relations

Many important groups (for example, simple or periodic) have no presentations with a $C'(\lambda)$-condition for small λ or with similar conditions. But by developing the van Kampen-Lyndon geometric method, one can construct groups with completely new properties. We formulate one of the first applications of graded diagrams to illustrate some new ideas concerning maps and diagrams. The statements of other new theorems will be presented in Sections 5-11 below.

Theorem 1 (A Yu Ol'shanskii [7]). *For every sufficiently large prime* p (e.g. $p > 10^{75}$) *there exists an infinite group* G *such that every proper subgroup of* G *has order* p.

This example combines two strong finiteness conditions: it is a quasi-finite group and a group of bounded exponent (G is a quasi-finite group if it is infinite but every proper subgroup of G is finite). This theorem answers a number of well-known questions on groups with the minimum condition for subgroups, and groups with maximum conditions [8, 1]. The group G is a Tarski monster (that is, an infinite group in which every proper subgroup has the same prime order). G is presented effectively and there exist equality and conjugacy algorithms for G.

To construct a presentation

$$G = \langle a_1, a_2 \mid r_1 = 1, r_2 = 1, \dots \rangle$$

of G, one has to define words $r_1, r_2, \dots$ of two kinds: words of the first kind look like A^p, where A is a cyclically reduced word in the alphabet $\{a_1^{\pm 1}, a_2^{\pm 1}\}$, and words of the second kind are more complicated.

We first concentrate our attention on the relators A^p, and formulate a precise presentation of another group, the *free Burnside group* **B**(m,n) of odd exponent $n > 10^{10}$. (We may also call it the Novikov-Adyan group [9, 10].) We use **B**(m,n) as an important example because its relators are easily described. The following definition of a presentation having an independent set of defining relations is based on the induction-like definitions in the papers of P S Novikov and S I Adyan for odd $n \geq 665$ [9, 10]. But in contrast to these papers, the inductive definition in [11] of an independent set of defining relations for the group **B**(m,n) is very straightforward.

Let $A_1 = a_1$ and

$$G(1) = < a_1, ..., a_m \mid A_1^n = 1 >.$$

For $i > 1$ we define A_i to be (one of) the shortest of those words whose images have infinite order in the group

$$G(i-1) = < a_1, ..., a_m \mid A_1^n = 1, ..., A_{i-1}^n = 1 >.$$

It has been shown [11] that $\{A_i^n\}_{i=1}^{\infty}$ is an independent system of relators for **B**(m,n).

It is clear that the presentation

$$G = < a_1, ..., a_m \mid A_1^n = 1, ... >, \quad m \geq 2 \tag{2.1}$$

gives us a periodic group, but it is not clear why G is infinite. It is impossible to use any $C'(\lambda)$-condition (considered above) for proving the group G to be infinite. Indeed, let $r_1 = a_1^n$. Then for a large number i we get $r_i = (a_1^k a_2)^n$, where $k = [n/2]$, and the piece a_1^k has length k, where $k \approx \frac{1}{2} |r_1|$. So we need new conditions.

An important technical tool in dealing with relations of the type $A^n = 1$ is the use of diagrams with long periodic words written on their boundaries and on the boundaries of their regions (Fig. 4).

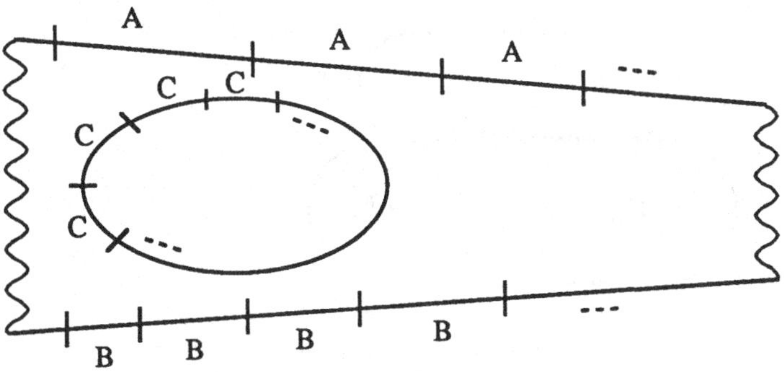

Fig. 4

A *periodic word* with period A is a subword of some power A^{ℓ} of a word A. For example,

$$X \equiv a^2ba^2ba^2$$

is a periodic word with period a^2b or aba or ba^2 (cyclic conjugates of a^2b).

$$A \equiv a^2b$$

is a simple period and

$$A' \equiv a^2ba^2b$$

is not a simple period since $A' \equiv A^2$. Words a^3ba^2b and $a^2ba^2ba^2b^{-1}$ are also simple periods for X.

It is easy to prove that if X is both A-periodic and B-periodic (where A and B are simple) then either B is a cyclic conjugate of A or

$$|X| < |A| + |B|.$$

This property of periodic words is useful in our proofs as well as in the proofs of P S Novikov and S I Adyan. In particular, if $|X| >> |A|$ and B is not a cyclic conjugate of A, then $|X| < |B|\,(1+\varepsilon)$, where ε is a small real value.

Let p be an arc in a reduced diagram Δ between two regions Π_1, Π_2 with boundary labels A^n and B^n and $\phi(p) = X$, Fig. 5.

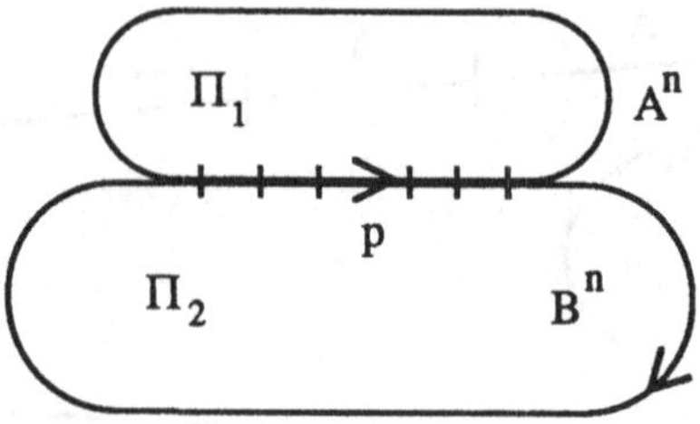

Fig. 5

Then we can rewrite the above condition

$$|p| < (1+\varepsilon)\,|B| << |B^n| = |\partial\Pi_2|,$$

because n is large and Δ is a reduced diagram (so that B is not a cyclic conjugate of A). Hence we have to analyse maps with conditions which are similar to the following:

$$|p| > \varepsilon|\partial\Pi_1| \Rightarrow |p| < \varepsilon|\partial\Pi_2|$$

or

$$\min(\,|p|\,/\,|\partial\Pi_1|,\ |p|\,/\,|\partial\Pi_2|\,) \le \varepsilon. \tag{2.2}$$

Let us try to imagine a map satisfying (2.2) which is composed of two kinds of region: "small" and "large". Certainly, our imagination suggests a picture like Fig. 6.

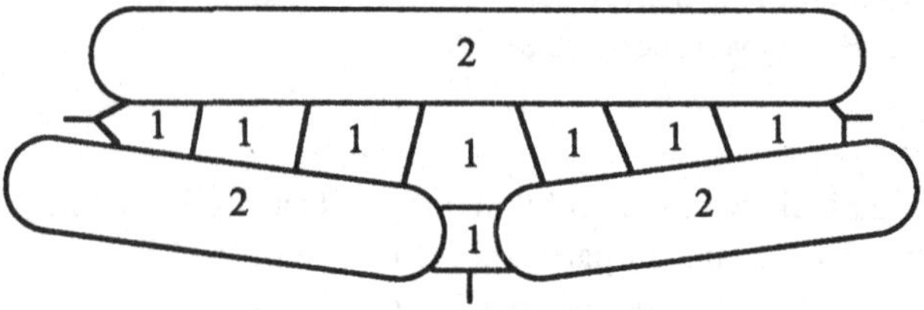

Fig. 6

Small regions are compressed between large ones and form long narrow strips. So we have to analyse long strips and must replace the usual conditions of small cancellation theory by certain new conditions.

3. Contiguity submaps and the A-condition

The $C'(\lambda)$-condition has to do with arcs between regions. The A-condition, which was introduced in [1, 11] concerns long narrow strips between regions. The idea of closeness of regions is formalized in the notions of graded map and contiguity submap.

By *a graded map* we mean a map M and a rank-function on the set M(2) of regions:

$$r(\Pi) = i, \ \ \Pi \in M(2), \ i = 0, 1, 2, \dots$$

$$r(M) = \max_{\Pi \in M(2)} r(\Pi).$$

The notion of *contiguity submap* is defined by induction. A 0-contiguity submap Γ between regions Π and Π' looks like Fig. 7.

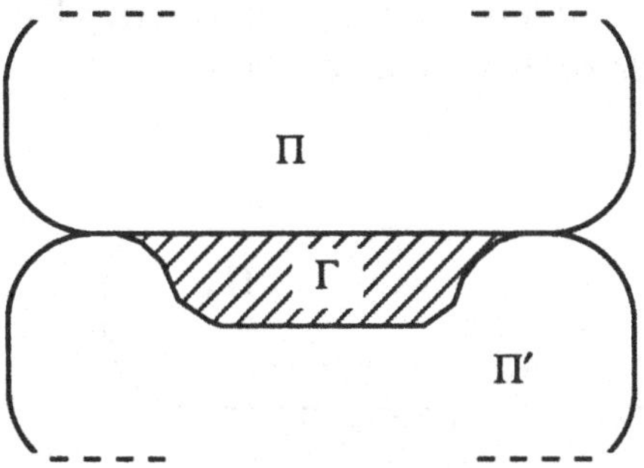

Fig. 7

A k-contiguity submap Δ $(k > 0)$ looks like Fig. 8.

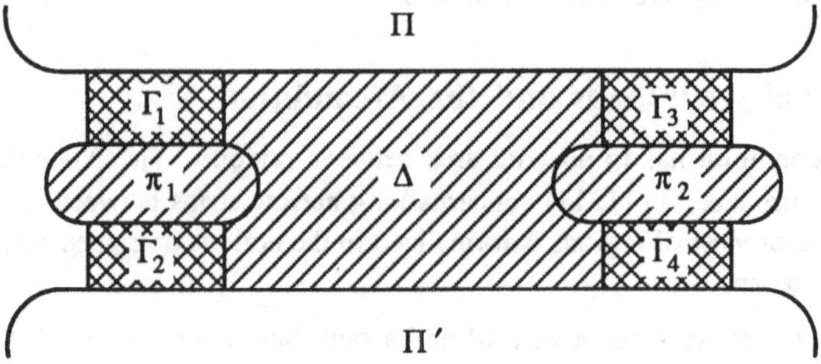

Fig. 8

In Fig. 8

$$\max(r(\pi_1), r(\pi_2)) = k,$$

and Γ_j is a k_j-contiguity submap, where $k_j < k$ and $\partial\Gamma_j$ is long compared to $\partial\pi_1$ and $\partial\pi_2$, $j = 1,2,3,4$. (For instance, $|\partial\Gamma_1| > \varepsilon|\partial\pi_1|$, etc. One can find a more precise definition in [1, 11].)

Let Δ be a contiguity submap with boundary path $p_1q_1p_2q_2$, where q_1, q_2 are subpaths of the boundary paths of Π_1 and Π_2, see Fig. 9.

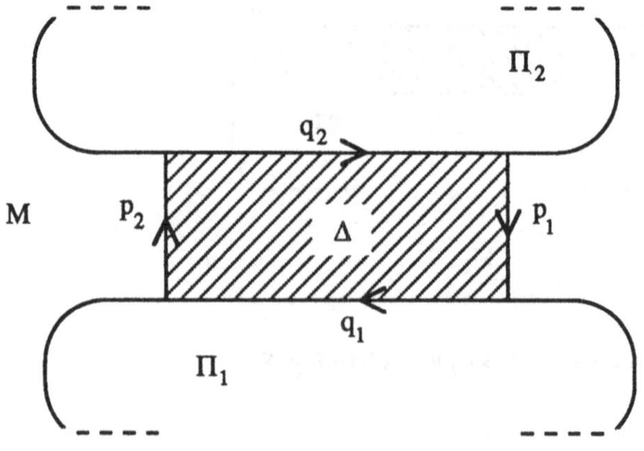

Fig. 9

Then the ratio

$$|q_1| / |\partial\Pi_1| =: (\Pi_1, \Delta, \Pi_2)$$

is called the *degree of contiguity* of Π_1 to Π_2 via the submap Δ.

We can now formulate the A-*condition*. Let n be a rather large number (say $n > 10^{10}$). We say that a graded map M is an A-*map* if

A1. $|\partial\Pi| \geq nr(\Pi)$ for any region $\Pi \in M(2)$;

A2. if q is a subpath of the boundary path of $\Pi \in M(2)$ and $|q| \leq r(\Pi)$ then q is a geodesic path in M;

A3. if Δ is a contiguity submap in M (Fig. 9) and $(\Pi_1, \Delta, \Pi_2) \geq \varepsilon$, then

$$|q_2| \leq (1+\varepsilon)\, r(\Pi_2)$$

(ε is a small value, for instance, $\varepsilon = 0.001$).

Condition A3 generalizes the property of periodic words mentioned above. But we would like to emphasize that the A-condition concerns only maps (not diagrams). There are no label-functions or relations in groups here.

The properties of graded A-maps are similar to those of $C'(\lambda)$-maps. They can be proved by induction on the number of regions. We have the following result.

Lemma. *Let* M *be an* A-*map and* $r(M) > 0$. *Then there exists a region* Π *in* M *and a contiguity submap* Γ *such that* $(\Pi, \Gamma, \partial M) > 0.9$. *Moreover, there exists a region* π *in* M *such that* $\partial\pi$ *and* ∂M *contain a common arc* q *and* $|q| > 1/3\ |\partial\pi|$, (see Fig. 10).

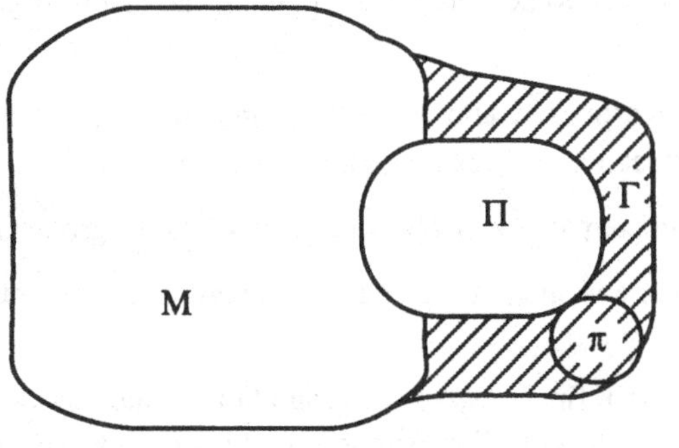

Fig. 10

We call such a region π an *external region.*

A number of other properties of A-maps have been proved. As a matter of fact we can say that an A-map looks like a "thick tree" or a "cactus", see Fig. 11.

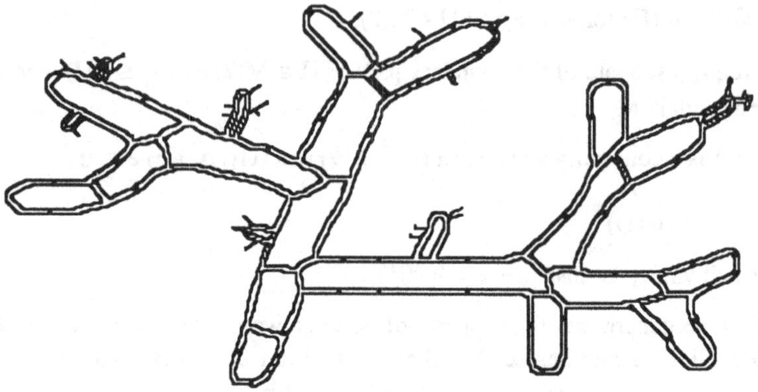

Fig. 11

4. Graded presentations

Now let us return to the presentation (2.1). We can make it into a graded presentation by declaring that a relator $r_i = A_i^n$ has rank $|A_i|$. Then every diagram over G is a graded diagram (with $r(\Pi) = |A|$ if the boundary label of Π is A^n).

A graded diagram is said to be *reduced* if it contains no pair of regions Π_1, Π_2 of rank $i \geq 1$ such that

(1) starting at vertices u and v on $\partial\Pi_1$ and $\partial\Pi_2$ in the clockwise and counterclockwise directions we read the same boundary labels;

(2) there is a simple path u-v such that $\phi(u\text{-}v) = 1$ in the group $G(i\text{-}1)$.

It is easy to see that any graded van Kampen diagram can be transformed to a reduced one.

We have to prove the infiniteness of the group G presented by (2.1) for proving the Novikov-Adyan theorem. First suppose that every reduced graded diagram over G is an A-map.

It is well-known that there exists an infinite set of semigroup words

$$w_k = w_k(a_1, a_2)$$

without subwords X^3, where X is a non-empty word. We shall explain the infiniteness of the group G if we can prove that $w_k \neq w_j$ in G for $j \neq k$.

Assume that $w_k w_j^{-1} = 1$ in G. Then there exists a reduced graded van Kampen diagram M over G with boundary label $w_k w_j^{-1}$. By the lemma in Section 3 there exists an external region π with an external arc q such that

$$|q| > 1/3\, |\partial\pi| = n/3\, |A|,$$

where A is a period of rank $|A|$. The label $\phi(q)$ is an A-periodic word. It contains a subword A^t (and so $A^{[t/3]}$ is a subword of w_k or w_j), where

$$t \approx [n/3] >> 3,$$

a contradiction to the choice of the words w_k, w_j.

Hence, it is necessary to prove that any graded diagram over G is an A-map. It is sufficient, of course, to examine this property for every G(i), where i is now an inductive parameter.

The main condition A3 concerns contiguity submaps. Let Δ be a contiguity submap of a diagram M, where $r(M) = i$, Fig. 9. Then

$$r(\Delta) < r(\Pi_1),\ r(\Delta) < r(\Pi_2).$$

Hence $r(\Delta) < i$ and, by the inductive hypothesis, Δ is an A-map. So the map Δ looks like a tree. On the other hand subpaths q_1, q_2 of boundary paths of regions are rather smooth. So Δ is a "long and narrow" diagram.

Therefore, a new problem arises: to investigate long narrow diagrams, with boundary $p_1q_1p_2q_2$, where

$$|q_1|, |q_2| >> |p_1|, |p_2|$$

and $\phi(q_1)$, $\phi(q_2)$ are periodic words with some periods A and B^{-1}, see Fig. 12 below.

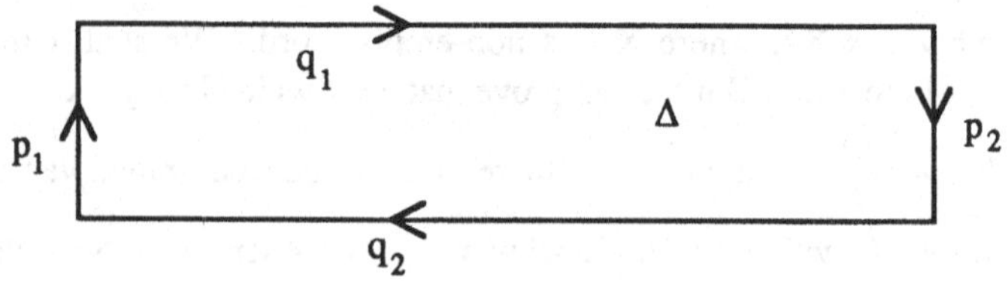

Fig. 12

Recall that in the simple case $|p_1| = |p_2| = 0$ and $r(\Delta) = 0$, we arrived (as in Section 2) at two possibilities:

(1) $A \equiv B$ or

(2) if $|q_1| >> |A|$, then $|q_2| < (1+\varepsilon)|B|$.

It turns out that a similar property can be extended by induction to long narrow diagrams over G(i-1). We can reject the first alternative because the diagram M, Fig. 9, is not reduced if A is a period for both regions Π_1 and Π_2. The second alternative yields condition A3.

Of course, the above reasoning is just an outline. Nevertheless the full proof of the infinity of **B**(m,n) is much shorter in [1, 11] than the original proof in [9, 10].

Let us return to Theorem 1 on the Tarski monster.

There are two kinds of relations, and relations of the second type look like

$$a_1BA^nBA^{n+2} \dots BA^{n+2k} = 1,$$

or

$$a_2BA^{n+1}BA^{n+3} \dots BA^{n+2k+1} = 1$$

where n, k are large numbers, a_1, a_2 are generators, and A, B are words in the alphabet $\{a_1^{\pm 1}, a_2^{\pm 1}\}$. These relations ensure that the subgroup $< A,B >$ contains a_1, a_2. Thus $< A,B > = G$. Such relations of all ranks guarantee that if $AB \neq BA$ in G, then $G = < A,B >$. So every proper subgroup of the group G is abelian. Moreover, if $AB = BA$, then

$$A = C^k, B = C^\ell \text{ for some } C \in G.$$

(This property is proved using the fact that any diagram over G on the surface of a torus is not reduced.) So every proper subgroup of G is cyclic.

In this case the relator r_i is not periodic, although it contains a lot of long periodic subwords. So we have to investigate contiguity submaps between separate sections of boundary paths of regions. This gives rise to the notion of B-map and many other details, which we omit.

5. Varieties of groups

5.1. Let $\mathfrak{B}_n$ be the *Burnside variety* defined by the law $x^n = 1$. The Novikov-Adyan method is a useful means of studying the free groups **B**(m,n) in the variety $\mathfrak{B}_n$ for odd $n \geq 665$. So also is the A-map method [1] for odd $n >> 1$. Both approaches are applicable to varieties defined by laws of the type

$$v(x_1,...,x_m)^n = 1,$$

where v is a given word. For instance, the independent system of laws

$$[x^{pn}, y^{pn}]^n = 1, p = 2, 3, 5, 7, 11, ...$$

has been considered by S I Adyan [10, 12], and the first author [13] has found a 2-generator periodic simple group of unbounded exponent in which the law $[x,y]^n = 1$ holds.

But, of course, many problems in the theory of varieties cannot be solved by using laws of this kind; more complicated ones are needed. For example, the following question was raised by H Neumann in her book [14] more than 20 years ago. (Recall that a variety of groups $\mathfrak{V}$ is called non-abelian if it contains a non-abelian group). Does there exist a non-abelian variety in which all the finite groups are abelian? This problem was solved in the affirmative [1, 15] by the second author.

Let $[x,y] := xyx^{-1}y^{-1}$ be a commutator,

$$v(x,y) := [[x^m, y^m]^m, [y^m, x^{-m}]^m]$$

and

$$w(x,y) := [x,y]\, v(x,y)^n [x,y]^{\varepsilon_1} v(x,y)^{n+1} \dots [x,y]^{\varepsilon_k} v(x,y)^{n+k}, \tag{5.1}$$

where m, n, k are sufficiently large numbers, $\varepsilon_i = \pm 1$, and $\varepsilon_1, ..., \varepsilon_k$ are fixed and satisfy $\sum_{i=1}^{k} \varepsilon_i = 0$.

It is easy to see that if a group G is finite and $w(x,y) = 1$ is a law in G, then G is abelian. Indeed, let us assume that $G' = [G,G] \neq 1$. By the Miller-Moreno Theorem, G contains a non-abelian metabelian subgroup H. Therefore $v(x,y) = 1$ is a law in H and $w(x,y) = [x,y]$ in H since $\sum \varepsilon_i = 0$. So $[x,y] = 1$ in H, a contradiction.

Theorem 2 (A Yu Ol'shanskii [1, 15]). *The law* $w(x,y) = 1$, *see* (5.1), *defines a non-abelian variety* $\mathcal{V}$ *having no non-abelian finite groups.*

To prove the theorem we need to construct the m-generator free group $F = F(m,\mathcal{V})$ in the variety $\mathcal{V}$ and then prove that $F' \neq 1$ for $m \geq 2$.

The outline of the construction of

$$G = \langle a_1 \dots a_m \mid r_1 = 1, \dots \rangle$$

is similar to that in Theorem 1 but the proof is more complicated because it is more difficult to satisfy the law $w(x,y) = 1$.

A M Storozhev and O N Matsedonska noted that $\mathcal{V}$ is the first example of a non-abelian variety whose groups possess the T-property (i.e. normality of subgroups is a transitive relation).

5.2. Theorem 2 has recently been strengthened by the first author as follows.

Let $\beta = \gamma_1\gamma_2 \dots$ be an infinite word in the alphabet $\mathbf{O} = \{0,1\}$, $\gamma_k \in \mathbf{O}$, let

$$m_k = (p_1 \dots p_k)^k,$$

where p_k is the kth prime number, $k \in N$, and

$$v_k(x,y) = [x^{m_k}, [x,y]^{n_0}], \tag{5.2}$$

$$w_k(\beta)(x,y) =$$

$$[x,y]v_k^{n_k+1}[x,y]v_k^{n_k+2}[x,y]^{-1}v_k^{n_k+3}[x,y]^{-1}\, v_k^{n_k+4} \dots [x,y]v_k^{n_k+4h-3}[x,y]v_k^{n_k+4h-2}$$

$$[x,y]^{-1}v_k^{n_k+4h-1}[x,y]^{-1}\, v_k^{n_k+4h}[x,y]^{\gamma_k}. \tag{5.3}$$

where h and the n_k, $k = 0,1,\dots$, are to be specified later.

Then the variety $\mathcal{V}_\beta$ is defined by the laws $w_k(\beta) = 1$, $k \in N$.

If $\beta \neq \beta'$, then there is a k such that $\gamma_k \neq \gamma'_k$. Comparing $w_k(\beta)$ with $w_k(\beta')$ (see (5.3)), we have that the laws $w_k(\beta) = 1$, $w_k(\beta') = 1$ imply the law $[x,y] = 1$. So

$$\mathcal{V}_b \underset{b \neq b'}{\cap} \mathcal{V}_{b'} = \mathcal{A}$$

where $\mathcal{A}$ is the variety of all abelian groups.

Let us suppose that the infinite word β contains infinitely many 1's. Then given a number n, there is a k such that n divides m_k and $\gamma_k = 1$. Therefore, if $a,b \in G$, $G \in \mathcal{V}_\beta$, and $[a^n,b] = 1$, then by (5.3) we have $w_k(\beta)(a,b) = 1$, whence by (5.2), the equality $w_k(\beta)(a,b) = 1$ implies the equality $[a,b] = 1$. Thus, in $\mathcal{V}_\beta$, the quasi-laws

$$[x^n,y] = 1 \Rightarrow [x,y] = 1$$

hold. In particular, in $G \in \mathcal{V}_\beta$, any elements of finite order lie in the centre of G and all torsion groups in $\mathcal{V}_\beta$ are abelain.

Now let h, n_0, n_1, n_2, ... in (5.2), (5.3) be such that

$$h = 10^5,\ n_0 >> 1,\ n_k >> n_{k-1},\ k \in N.$$

Then for any infinite word β a 2-generator non-abelian group $G(\beta)$ has been constructed in which the laws $w_k(\beta) = 1$, hold for all $k \in N$. Since the infinite words in **O** containing infinitely many 1's are continuously many, we have the following.

Theorem 3 (S V Ivanov [16, 17]). *There exist continuously many non-abelian varieties* $\mathcal{V}_\beta$ *defined as above such that:*

(1) $\mathcal{V}_\beta \cap \mathcal{V}_{\beta'} = \mathcal{A}$ for $\beta \neq \beta'$;

(2) *if* $G \in \mathcal{V}_\beta$, $a,b \in G$, $k \in N$, *and* $[a^k, b] = 1$, *then* $[a,b] = 1$; *in particular, all torsion groups in* $\mathcal{V}_\beta$ *are abelian.*

Part (2) of the theorem answers affirmatively a question of A L Shmel'kin 4.73b [18] on the existence of non-abelian varieties in which all torsion groups are abelian.

In view of Zorn's lemma the following corollary of Theorem 3 is obvious.

Corollary 1. *The set of just-non-abelian varieties of groups is continuous.*

This gives important new information on the structure of the lattice of group varieties and answers the following three questions.

Question 1 (A Yu Ol'shanskii, 4.48b [18]). How many varieties are there that have no finite basis for their laws and in which any proper subvariety has a finite basis for its laws?

Question 2 (A V Kusnetsov, 6.16 [18]). How many varieties are there that contain only finitely or countably many subvarieties?

Question 3 (Yu G Kleiman, 8.20 [18]). How many varieties covering $\mathfrak{A}$ are there? ($\mathfrak{M}$ is said to cover $\mathfrak{V}$ if $\mathfrak{V} \subsetneq \mathfrak{M}$ and inclusions $\mathfrak{V} \subset \mathfrak{U} \subset \mathfrak{M}$ imply $\mathfrak{U} = \mathfrak{V}$ or $\mathfrak{U} = \mathfrak{M}$.)

We note that Yu G Kleiman [19] himself has found a continuous set of soluble varieties covering $\mathfrak{U}$ for a certain soluble variety $\mathfrak{U}$.

5.3. The question on the number of varieties covering $\mathfrak{V}$, where $\mathfrak{V}$ is atomic, i.e. a just-non-trivial variety, is even more natural and interesting. If $\mathfrak{V} = \mathfrak{A}_n$, where $\mathfrak{A}_n$ is the variety of all abelian groups of exponent n, n = 2,3,..., then it is easy to verify that the set of varieties covering $\mathfrak{V}$ is countable.

Theorem 4 (S V Ivanov [16, 17]). *There exist continuously many varieties* $\mathfrak{U}_\beta$ *such that:*

(1) $\mathfrak{U}_\beta$ *has prime exponent* $n >> 1$ (e.g. $n > 10^{216}$),

(2) $\mathfrak{U}_\beta \cap \mathfrak{U}_{\beta'} = \mathfrak{A}_n$, *for* $\beta \neq \beta'$.

Corollary 2. *For any prime* $n >> 1$ *the set of varieties covering* $\mathfrak{A}_n$ *is continuous.*

The corollary gives answers to the following questions.

Question 4 (A V Kuznetsov, 6.16 [18]). How many varieties are there that are not locally finite and in which all proper subvarieties are locally finite (i.e. consist of locally finite groups)?

Question 5 (R A Bairamov, 2.1 [20]). Let a semigroup variety $\mathfrak{M}$ contain a finite number of semigroup subvarieties only. Is $\mathfrak{M}$ then locally finite? (This is true if $\mathfrak{M}$ is generated by its finite semigroups.)

Question 6 (R A Bairamov, 2.3 [20]). For $\mathfrak{M}$ as in Question 5, can the set of semigroup varieties covering $\mathfrak{M}$ be continuous?

The varieties $\mathcal{U}_\beta$ of Theorem 4 are constructed as follows. As above, let $\mathbf{O}$ be the alphabet $\{0,1\}$, let $W_1, W_2, \ldots$ be words in $\mathbf{O}$ having no non-empty subword E^3,

$$|W_1| >> 1, |W_k| = |W_{k-1}| + 8,$$

and let

$$V_k \equiv W_a\, 000\, W_{a+1}\, 000 \ldots W_{a+h-1}\, 000,$$

where $a = (k-1)h + 1$, $h = 10^6$, $k \in \mathbf{N}$.

Let $\beta = \gamma_1\gamma_2 \ldots$ be an infinite word in $\mathbf{O}$, let the word V_k be $\delta_1 \ldots \delta_\ell$, where $\delta_j \in \mathbf{O}$, and let

$$u(x,y) = [[x,y]^m, x],$$

where m is the integer part of $n^{1/3}$, for some prime $n >> 1$.

Then the word $w_{V_k}(\beta)(x,y)$ is defined by the formula

$$w_{V_k}(\beta)(x,y) = [x,y]u^{m+\delta_1}[x,y]u^{m+\delta_2}[x,y]^{-1}\, u^{m+\delta_3}[x,y]^{-1}u^{m+\delta_4} \ldots$$

$$[x,y]u^{m+\delta_{\ell-3}}[x,y]u^{m+\delta_{\ell-2}}[x,y]^{-1}u^{m+\delta_{\ell-1}}[x,y]^{-1}u^{m+\delta_\ell}[x,y]^{\gamma_k}. \qquad (5.4)$$

Then the variety $\mathcal{U}_\beta$ is defined by the laws

$$x^n = 1,\ w_{V_k}(\beta)(x,y) = 1,\ k \in \mathbf{N}.$$

As in Section 5.2, if $\beta \neq \beta'$ then comparing the words $w_{V_k}(\beta)$ with $w_{V_k}(\beta')$, see (5.4), for some k gives that $\mathcal{U}_\beta \cap \mathcal{U}_{\beta'} = \mathcal{A}_n$.

For any β the non-abelian 2-generator group $G(\beta)$ has been constructed. The construction of $G(\beta)$ is rather complicated even if compared with the proofs of Theorems 2, 3.

Moreover, the independence of the laws $w_{V_k}(\beta) = 1$, $k \in \mathbf{N}$, has been established in $\mathbf{B}(2,n)$. Hence, $\mathbf{B}(2,n)$ does not satisfy the minimum or maximum condition for verbal subgroups.

We note that S I Adyan [10, 21] and V S Atabekyan [22] have shown that, for $m > 1$, and odd $n >> 1$, $\mathbf{B}(m,n)$ does not satisfy the maximum or minimum condition for normal subgroups. We also mention that A M Storozhev [23] constructed a non-abelian finitely based variety of large odd composite exponent in which all finite groups are abelian (for prime exponent this follows from Corollary 2).

5.4. A group G is called *hopfian* if any epimorphism $G \to G$ has trivial kernel.

G Baumslag [24] has posed the following problem. Is the group F/V(N) hopfian if the quotient group F/N is hopfian, where F is a free group of finite rank, N a normal subgroup of F, and V some set of group words?

H Neumann formulated Problem 15 [14] on the existence of nonhopfian relatively free groups of finite rank (that is, the most interesting and important case N = F in Baumslag's problem).

The hopficity of absolutely free groups and free polynilpotent groups of finite rank is well known and can be deduced from their residual finiteness. An example of a hopfian relatively free soluble group which is not residually finite was found by Yu G Kleiman [26].

The following theorem presents the solution of the problems of G Baumslag and H Neumann.

Theorem 5 (S V Ivanov [16, 25]). *There exists a group variety $\mathfrak{M}$ in which all relatively free groups of rank more than one are nonhopfian.*

Let us explain the construction of $\mathfrak{M}$ and how the nonhopficity arises.

Let $U = [x^n, y^n]$, where $n >> 1$,

$$V(x,y) = U^n y U^{n+1} y U^{n+2} y^{-1} U^{n+3} y^{-1} U^{n+4} y U^{n+5} y^{-1}, \tag{5.5}$$

and $W(x,y) = xV(x,y)$. Then the words $w_1(x,y)$, $w_2(x,y)$ are given by the following formulae:

$$w_1(x,y) = [(W^n y)^2, y^{-1}][(W^{n+1} y)^2, y^{-1}] \dots [(W^{n+h} y)^2, y^{-1}]V \tag{5.6}$$

$$w_2(x,y) = [(W^{2n} y)^2, y^{-1}][(W^{2n+1} y)^2, y^{-1}] \dots [(W^{2n+h} y)^2, y^{-1}], \tag{5.7}$$

where $h = 10^6$.

The variety $\mathfrak{M}$ is defined by the laws $w_1(x,y) = 1$, $w_2(x,y) = 1$.

Let $H = F_2(\mathfrak{M}) = \langle a,b \rangle$ be the free group of rank two in $\mathfrak{M}$, and let c denote the element of H which is equal to V(a,b) (see (5.5)). In view of the equality $w_1(a,b) = 1$ in H and formula (5.6), we have $c \in \langle ac,b \rangle \subset H$, whence $H = \langle ac,b \rangle$.

We now put

$$w_3(x,y) = [(x^{2n} y)^2, y^{-1}][(x^{2n+1} y)^2, y^{-1}] \dots [(x^{2n+h} y)^2, y^{-1}]. \tag{5.8}$$

Since

$$w_3(W(x,y), y) = w_2(x,y)$$

(see (5.7), (5.8)), we have by definition $w_3(ac,b) = 1$. Hence, if ac, b are free generators of the group H, i.e. the epimorphism $\alpha : H \twoheadrightarrow H$, where

$$\alpha(a) = ac, \alpha(b) = b,$$

is an isomorphism, then $w_3(a,b) = 1$.

With the help of the construction of a graded presentation for H, it has been proved that if r(x,y) is a cyclically reduced non-empty word and r(a,b) = 1 in H, then

$$|r(x,y)| > 1/2\ |w_1(x,y)| > n^3.$$

On the other hand, $|w_3(x,y)| < 5hn$. Since $n >> h$, $w_3(a,b) \neq 1$ in H and ac, b are not free generators of H, i.e. H is not hopfian.

It is worth noting that in the proof, the following property of the word V(x,y) is crucial. If A_1, B_1, A_2, B_2 are words in the alphabet $\{a^{\pm 1}, b^{\pm 1}\}$ and $V(A_1, B_1)$ is conjugate to $V(A_2, B_2)$ in H, then there exists a word C such that the equalities

$$A_1 = CA_2C^{-1}, B_1 = CB_2C^{-1}$$

hold in H. Words V(x,y) with this property are very hard to find even in the free group $F_2 = < a,b >$, and a word of the kind in (5.5) is the simplest example of such a word known to the first author.

It remains to note that the nonhopficity of $F_m(\mathcal{M})$, for all $m > 1$, obviously follows from that of H.

5.5. The solution of another problem of H Neumann, Problem 19 in [14], is given by:

Theorem 6 (A M Storozhev [27]). *Let* $W_0 = V_0^{r+1}U_0^{\varepsilon_1}V_0^{r+2}U_0^{\varepsilon_2} \ldots U_0^{\varepsilon_h}V_0^{r+h}$,

where

$$V_0 = [[x^m, y^m]^n, [[y^m, z^m]^n, [z^m, x^m]^n]^n],$$

$$U_0 = [x^{10}, y^{10}]^{1000}.$$

Then for suitable values of the parameters $\varepsilon_1, \ldots, \varepsilon_h \in \{\pm 1\}$, $m,n,r \in \mathbb{N}$, *the verbal subgroups* $W_0(F_2)$, $W_0(F_3)$ *of the free groups*

$$F_2 = < a_1,a_2 >, F_3 = < a_1, a_2, a_3 >$$

are such that the shortest non-trivial element in $W_0(F_2)$ *is longer than the shortest one in* $W_0(F_3)$, *which is equal to* $W_0(a_1, a_2, a_3)$.

6. Extensions, relation modules and relation bimodules of certain aspherical groups

6.1. Let a group G be given by a graded presentation

$$< \mathfrak{A} \mid \mathcal{R} >, \text{ where } \mathcal{R} = \bigcup_{i=1}^{\infty} \mathcal{R}_i. \tag{6.1}$$

This presentation is called *aspherical* if there exist no reduced graded diagrams (containing $\mathcal{R}$-cells) over G on the surface of a sphere. Let us suppose that $\mathcal{R}$ is not symmetrized and contains no cyclic conjugate $\bar{r}$ of any $r \in \mathcal{R}$ such that $\bar{r} \not\equiv r$.

In the case $\mathcal{R} = \mathcal{R}_1$ (when (6.1) is not graded) the following property of an aspherical presentation is well known [3]. Let G = F/N, where

$$F = < \mathfrak{A} \mid \varnothing >,$$

N is the normal closure of $\mathcal{R}$ in F and F/[F,N] is the free central extension of G. Then the quotient group N/[F,N] is a free abelian group with free basis

$$\{r^c \mid r \in \mathcal{R}\},$$

where r^c is the coset r[F,N].

Of course, it is not surprising that the same is true in the graded case. For instance, the Schur multiplicator of the free Burnside group **B**(m,n), for $m > 1$ and odd $n >> 1$, is a free abelian group of infinite rank, see [28].

Let us consider the free central extension **C**(m,n) of **B**(m,n). It is easy to prove that the m-generator free group **C**(m,n) in the variety given by the law $[x^n,y] = 1$ is torsion-free and

$$Z = \mathbf{C}^n(m,n) = < c^n \mid c \in \mathbf{C}(m,n) >$$

is a free abelian group of infinite rank equal to the centre of **C**(m,n), and

$$\mathbf{B}(m,n) = \mathbf{C}(m,n) / Z,$$

see [28].

So it is possible to choose any subgroup $H \subset Z$ and to obtain various central extensions **C**(m,n) / H of the group **B**(m,n). In particular, we can get in this way the Adyan groups **A**(m,n) [29, 10] without torsion in which $K \cap M \neq 1$ for any pair of its non-trivial subgroups K, M.

Another example has been noted by the second author [30]. It is possible to choose $H \subset Z$ such that $|Z/H| = n$ and for every non-central element

$$g \in L = C(m,n) / H,$$

we have $g^n \in C = Z / H$, $g^n \neq 1$ in L. So if $C = \{c_1, ..., c_n\}$, where $c_1 = 1$ and $M_1, ..., M_n$ are the sets of solutions of the equations $x^n = c_1, ..., x^n = c_n$ in L, then in the group L with any topology the sets $M_1, ..., M_n$ are closed. Since

$$\{1\} = L \setminus (M_2 \cup M_3 \cup ... \cup M_n \cup \{c_2, ..., c_n\}),$$

$\{1\}$ is an open set. Consequently, L admits the discrete topology only. This gives a solution of Markov's problem [31] on the existence of a countable group which admits the discrete topology only. An uncountable example of such a group was given by S Shelah [32].

The following theorem has been proved with the help of another quotient group of a uncountable free group with the law $[x^n, y] = 1$.

Theorem 7 (I N Zyabrev, E A Reznichenko [84]). *There exists a connected topological group* G *such that* G *has no exponent* n *while* $x^n = 1$ *holds in a certain neighbourhood of the identity.*

This theorem answers Platonov's question 4.11 [18] on the existence of a topological connected group with a non-trivial law which holds in a neighbourhood of the identity but does not hold in the whole group.

A certain central extension of the Tarski monster from Theorem 1 gives us:

Theorem 8 (I S Ashmanov, A Yu Ol'shanskii [28]). *There exists a non-abelian group* G *such that every proper subgroup of* G *is infinite cyclic and* $H \cap K \neq 1$ *for any pair of proper subgroups* $H, K \subset G$.

Central extensions of other aspherical groups turn out to be useful to disprove the conjecture of P Hall [33] that if a word $v(x_1, ..., x_m)$ has a finite set of values in a group G, then the verbal subgroup v(G) of G is finite.

Theorem 9 (S V Ivanov [13], A Yu Ol'shanskii [1]). *There is a group* G, *a word* v(x,y) *and an element* $g \in G$ *such that* g *has infinite order while the set of values of the word* v(x,y) *in the group* G *is* $\{1,g\}$.

The first author proposed a central extension of a 2-generator torsion group as G and the word

$$[[x^{pn}, y^{pn}]^n, y^{pn}]^n$$

as v(x,y), where n is odd, p is prime, $n >> p > 5000$. The second author has proposed a central extension of a 2-generator free group in the variety of Theorem 2 and the word (5.1).

6.2. Let us pass on to free abelian extensions of aspherical groups. As above, let G = F/N, and let $\psi : \mathbf{Z}(F) \twoheadrightarrow \mathbf{Z}(G)$ be the natural epimorphism of integral group algebras.

In the ungraded case $\mathfrak{R} = \mathfrak{R}_1$ the following theorem can be found in the book of Lyndon-Schupp [3] and a revised proof in the paper [34] of D Collins and J Huebschmann.

Theorem 10 (I S Ashmanov, A Yu Ol'shanskii [28]). *Let the graded presentation* (6.1) *of the group* G *be aspherical and let* q(r) *be a word such that*

$$(q(r))^k = r$$

and q(r) *is not a proper power in* F. *Then the (left) relation* $\mathbf{Z}(G)$*-module* M = N/[N,N] *of the group* G = F/N *is generated by elements* r^a, $r \in \mathfrak{R}$, *where* $r^a = r[N,N]$, *and has the following defining relations:*

$$(1 - q(r))^{\psi} \circ r^a = 0, r \in \mathfrak{R}.$$

In the paper [35] of the first author the notions of relation module and relation bimodule for groups, semigroups and associative algebras given by defining relations are introduced. In particular, with the above notation, the relation bimodule B of the group G is defined as the abelian group R/R^2, where R = Ker ψ, with the following two-sided action of $\mathbf{Z}(G)$ on it:

$$f^{\psi} \circ t^b \circ g^{\psi} = (f \cdot t \cdot g)^b,$$

where $f,g \in \mathbf{Z}(F)$, $t \in R \subset \mathbf{Z}(F)$, $t^b = t + R^2$. (The notions of relation bimodules for semigroups and associative algebras are defined in a similar way.)

Theorem 11 (S V Ivanov [35]). *Under the conditions of Theorem 10, the relation bimodule* B *of the group* G = F/N *is generated by elements* r^b, $r \in \mathfrak{R}$, *where* $r^b = r + R^2$, *and has the following defining relations:*

$$(q(r))^{\psi} \circ r^b = r^b \circ (q(r))^{\psi}, r \in \mathfrak{R}.$$

In particular Theorems 10, 11 give obvious descriptions of the relation modules and bimodules of the groups **B**(m,n), for odd n >> 1, of the groups of Theorem 1, and others.

As has been noted, an ungraded presentation is a special case of a graded one. Therefore, Theorem 11 describes relation bimodules of 1-relator groups and small cancellation groups (a proof of the asphericity of these groups can be found in [3]).

For example, let $G = \langle \mathfrak{A} \mid q^k = 1 \rangle$ be a 1-relator group, q not a proper power. Then

$$B = \langle (q^k)^b \mid (q)^\Psi \circ (q^k)^b = (q^k)^b \circ (q)^\Psi \rangle.$$

In this connection we note that the description of a relation bimodule (and module) of a 1-relator semigroup quite recently obtained by the first author (based on his paper [36]) is more complicated.

7. Embedding theorems

It is known that small cancellation theory has been developed by R C Lyndon and others in the case of free products, see [3]. Similarly, the study of graded diagrams over free products is very fruitful.

More specifically, let **F** be a free product $\Pi^* G_\alpha$ of groups G_α, $\alpha \in I$, let

$$\mathfrak{A} = \cup G_\alpha$$

be the free amalgamation (union) of the groups G_α (then elements of **F** are considered as words in the alphabet $\mathfrak{A}$), let $\mathfrak{R} = \bigcup_{i=1}^{\infty} \mathfrak{R}_i$ be a graded set of defining words in the alphabet $\mathfrak{A}$, and let $N = \langle \mathfrak{R} \rangle^{\mathbf{F}}$ be the normal closure of $\mathfrak{R}$ in **F**. Then the quotient group G = **F**/N has by definition the following graded presentation:

$$G = \langle \mathfrak{A} \mid \mathfrak{R} \rangle.$$

With the help of defining relations similar to those appearing in the proof of Theorem 1 the following theorem has been proved.

Theorem 12 (V N Obraztsov [37]). *Let* $\mathfrak{A} = \cup G_\alpha$ *be a finite or countable free amalgamation of groups* G_α, $\alpha \in I$, *without involutions, and let* n *be odd,* $n >> 1$ (e.g. $n > 10^{78}$). *Then* $\mathfrak{A}$ *can be embedded in a 2-generator group* $G(\mathfrak{A})$ *in such a way that:*

(1) *any maximal subgroup* H *of* $G(\mathfrak{A})$ *is conjugate to* $G_\alpha \subset \mathfrak{A}$ *for some* α *or is cyclic of order* n;

(2) *if* H_1, H_2 *are distinct maximal subgroups of* $G(\mathfrak{A})$, *then* $H_1 \cap H_2 = 1$.

It is obvious that Theorem 12 generalizes Theorem 1 (take $\mathfrak{A} = 1$ and n prime) and the theorem due to the first author [38] on the existence of 2-generator groups of odd exponent $n >> 1$ whose maximal subgroups are all cyclic of order n ($\mathfrak{A} = 1$).

Theorem 12 can be modified for groups with involutions and any $n >> 1$ as follows.

Theorem 13 (A Yu Ol'shanskii [39]). *Let* $\mathfrak{A} = \cup G_\alpha$ *be a finite or countable free amalgamation of groups* G_α, $\alpha \in I$, $n >> 1$ (e.g. $n > 10^{78}$). *Then* $\mathfrak{A}$ *can be embedded in a 2-generator group* $G(\mathfrak{A})$ *in such a way that:*

(1) *any maximal subgroup* H *of* $G(\mathfrak{A})$ *is conjugate to* $G_\alpha \subset \mathfrak{A}$ *for some* α, *or is cyclic of order* n, *or is the infinite dihedral group;*

(2) *if* H_1, H_2 *are distinct maximal subgroups of* $G(\mathfrak{A})$, *then* $H_1 \cap H_2 = 1$, *or* $|H_1 \cap H_2| = 2$ *and at least one of* H_1, H_2 *is a dihedral group.*

In view of Theorem 12 the following two results hold.

Corollary 3. *Every finite or countable group* H *satisfying the law* $x^n = 1$ (*odd* $n >> 1$) *can be embedded in a 2-generator group* G *satisfying the same law* $x^n = 1$.

In fact, Corollary 3 asserts that the well-known embedding theorem due to G Higman, B Neumann, H Neumann [40] holds in the Burnside variety $\mathfrak{B}_n$.

Recall that a group G is called *quasi-finite* if all its proper subgroups are finite.

Corollary 4. *There exists a quasi-finite 2-generator group* G *such that every finite group* H *without involutions can be embedded into* G.

We can add to this.

Theorem 14 (G S Deryabina, A Yu Ol'shanskii [41]). *A finite group* G *can be embedded in a quasi-finite group if and only if* $G = G_1 \times G_2$, *where* G_1 *has odd order and* G_2 *is an abelian 2-group.*

Other corollaries of Theorem 12 indicated by V N Obraztsov are related to groups which satisfy the minimum condition for subgroups. A G Kurosh and S N Chernikov [42] posed the problem on the countability of every group with the minimum condition for subgroups. Obviously, if

$$G_\alpha \subset \mathfrak{A}, \alpha \in I,$$

are groups with the minimum condition for subgroups, then $G(\mathfrak{A})$ also satisfies the minimum condition. Hence we can iterate the embedding process using transfinite induction (up to the first uncountable ordinal) and obtain:

Corollary 5. *There exists a group* U *of odd exponent* $n >> 1$ *such that* U *satisfies the minimum condition for subgroups and has the first uncountable cardinality.*

An affirmative solution of the problem of S Shelah [32] on the existence of a countable simple group without maximal subgroups (uncountable groups of this type were constructed in [32]) also follows easily from Theorem 12 (and from another theorem due to the first author [38] which is not described here).

Corollary 6. *There exists a countable simple group* C *of odd exponent* $n >> 1$ *that has no maximal subgroups.*

Let us add that Corollary 6 is of interest because, according to [44], the existence of C means that the operations of forming the Frattini subgroup and the direct product do not commute, even in the class of countable groups.

8. Conjugacy relations and more embedding theorems

8.1. Recall that a group G is called *divisible* if for any $g \in G$, $n \in N$ the equation $x^n = g$ is soluble in G.

For a long time the well-known question of the existence of a finitely generated non-trivial divisible group has remained open. The breakthrough was made by:

Theorem 15 (V S Guba [45]). *There exists a 2-generator torsion free divisible group in which extraction of roots is unique, i.e.*

$$x^n = y^n \textit{ implies } x = y.$$

Simultaneously (in [45]), a question of D V Anosov (8.8a, [18]) has been answered by:

Theorem 16 (V S Guba [45]). *There exists a 2-generator group* G *containing an infinite cyclic subgroup* H *such that every element* g *of* G *is conjugate to an element of* H.

While in the proof of Theorem 1, periodic and fractional-periodic relators play the key role, in the paper [45] as well as in proofs of the theorems below, conjugacy relations of the kind $CAC^{-1} = B$, where A, B, C are specially chosen words (in particular, $|C| >> |A|, |B|$), are the most significant.

8.2. The first author has recently found an approach, simpler and more general than that in [45], to the definition of graded conjugacy relations and to the study of their consequences, and this is used in proving the remaining theorems in this section.

Theorem 17 (S V Ivanov [46,47]). *Let* G *be a finite or countable group without involutions, and let* G *contain an element of order* $n >> 1$ *(e.g.* $n > 10^{12}$ *or perhaps* $n = \infty$*). Then* G *can be embedded in a 2-generator (simple) group* S(G) *such that:*

(1) *every element of* S(G) *is conjugate to an element of the subgroup* $G \subset S(G)$, *i.e.* $S(G) = \cup\, sGs^{-1}$;

(2) G *is antinormal in* S(G), *i.e. for any* $s \notin G$, $sGs^{-1} \cap G = 1$.

It is easy to see that both restrictions in the theorem are essential: the theorem does not hold for groups of small exponent nor for finite 2-groups.

Theorems 15 and 16 are obvious corollaries of Theorem 17 (take $G = \mathbf{Q}$ and C_∞, where $\mathbf{Q}$ is the additive group of rational numbers and C_n is the cyclic group of order n).

Let us consider some other applications of Theorem 17.

First of all we note that in a p-group G any element $g \in G$ of order p is not conjugate with g^k if $g^k \neq g$. Therefore G contains at least p conjugacy classes.

Taking $G = C_p$ in Theorem 17, where p is a prime $>> 1$, we have the first example of a non-cyclic p-group with p conjugacy classes.

Corollary 7. *There exists a 2-generator infinite group of prime exponent* $p >> 1$ *which contains exactly* p *conjugacy classes.*

Taking $G = C_{p^\infty}$ in Theorem 17, where C_{p^∞} is a quasicyclic p-group, we obtain the first example of a non-trivial finitely generated divisible torsion group.

Corollary 8. *For any odd prime* p *there exists a non-trivial 2-generator divisible* p*-group.*

Let us point out that Corollaries 7, 8 were actually proved (by the first author) before Theorem 17, and appear in the book [1].

The following three theorems are well-known and are actually proved in [48, 40]. We give their formulations for the countable case only, and add some easily-deduced information on elements of finite order.

Theorem A (B H Neumann [48]). *Every finite or countable group* G *can be embedded in a countable group* A(G) *in which every element* a *of finite order is conjugate to an element of* G *or there is a prime* p, *and* k, $\ell > 0$ *such that* a^{p^k} *is conjugate to an element of* G *of order* p^ℓ. *In particular,* A(G) *has no involutions, if* G *has none.*

Theorem B (G Higman, B H Neumann, H Neumann [40]). *Every finite or countable group* G *can be embedded in a countable (mixed) group* B(G) *in which any two elements of the same order are conjugate and any element of finite order is conjugate to an element of* $G \subset B(G)$.

Theorem C (G Higman, B H Neumann, H Neumann [40]). *Every finite or countable group* G *can be embedded in a 2-generator group* C(G).

Theorem 17 allows us to combine Theorems A and C, and Theorems B and C, as follows.

Theorem 18 (S V Ivanov [46, 47]). *Every finite or countable group* G *without involutions can be embedded in a 2-generator divisible group* D(G) *in which any element* d *of finite order is conjugate to an element of* $G \subset D(G)$, *or there is a prime* p, *and* k, $\ell > 0$ *such that* d^{p^k} *is conjugate to an element of* G *of order* p^ℓ.

Theorem 19 (S V Ivanov [46, 47]). *Every finite or countable group* G *without involutions can be embedded in a 2-generator (mixed) group* E(G) *in which any two elements of the same order are conjugate and any element of finite order is conjugate to an element of* $G \subset E(G)$.

Theorems 18, 19 follow immediately from Theorems A and 17, B and 17 respectively, if we take

$$D(G) = S(A(G)),\ E(G) = S(B(G)).$$

The question of V S Guba 9.10 [18] on the existence of a finitely generated group of order more than two with exactly two conjugacy classes is answered by the following obvious corollary of Theorem 19.

Corollary 9. *Every countable torsion free group* G *can be embedded in a 2-generator torsionfree group* E(G) *whose non-trivial elements are all conjugate.*

Constructions of a free product with amalgamation and an HNN-extension used in the proofs of Theorems A and B lead to the appearance of elements of infinite order in enveloping groups. So the following two theorems have been proved separately for torsion groups.

Theorem 20 (S V Ivanov [46, 47]). *Every finite or countable torsion group* G *without involutions can be embedded in a 2-generator divisible torsion group* T(G) *in which any non-trivial element* t *is conjugate to an element of* $G \subset T(G)$ *or there is a prime* p *and* $k, \ell > 0$ *such that* t^{p^k} *is conjugate to an element of* G *of order* p^ℓ.

Theorem 21 (S V Ivanov). *Every finite or countable group* G *of prime exponent* $p >> 1$ (*e.g.* $p > 10^{12}$) *can be embedded in a 2-generator group* U(G) *of exponent* p *which contains exactly* p *conjugacy classes* (*and* $1, u, \ldots, u^{p-1}$ *are their representatives if* $u \neq 1$).

Let us note that in contrast to Corollary 7, it was not known whether any group of prime exponent $p >> 1$ could be embedded in some group (not finitely generated) of exponent p having p conjugacy classes. At the same time, it is obvious that any locally finite p-group with p conjugacy classes has order p.

We emphasize that the theorems of this paragraph deal with finitely generated enveloping groups, and torsion enveloping groups are not locally finite. It has long been known that among locally finite countable groups there exists a so-called universal group C (P Hall [50]). The group C is simple, contains every locally finite countable or finite group and any two isomorphic finite subgroups of C are conjugate (by the last two properties, C is defined up to isomorphism). In particular, C is divisible and any two elements of the same order are conjugate in C. At the same time, the first author [13] has noted that there exists no countable torsion group G such that every countable torsion group is isomorphic to a quotient group of a subgroup of G (i.e. to a section of G).

8.3. The proof of the following theorem strengthening Theorems 12, 17 is based on the application of both conjugacy relations and periodic and fractional-periodic relations.

Theorem 22 (S V Ivanov [46]). *Let* $\mathfrak{A} = \cup G_\alpha$ *be a free finite or countable amalgamation of groups* G_α, $\alpha \in I$, *without involutions, and let some* G_α *contain an element of order* $n >> 1$ (*e.g.* $n > 10^{99}$ *or perhaps* $n = \infty$). *Then* $\mathfrak{A}$ *can be embedded in a 2-generator group* $S(\mathfrak{A})$ *in such a way that:*

(1) *every maximal subgroup of* $S(\mathfrak{A})$ *is conjugate to* $G_\alpha \subset \mathfrak{A}$, *for some* α, *in particular, every element of* $S(\mathfrak{A})$ *is conjugate to an element of* $\mathfrak{A}$, *i.e.*

$$S(\mathfrak{A}) = \cup s\mathfrak{A}s^{-1};$$

(2) *if* $s\mathfrak{A}s^{-1} \cap \mathfrak{A} \neq 1$ *in* $S(\mathfrak{A})$, *then* $s \in G_\alpha$ *for some* α *and*

$$s\mathfrak{A}s^{-1} \cap \mathfrak{A} = G_\alpha$$

or $s = 1$.

This theorem establishes the existence of groups with interesting new properties. Thus, by taking

$$\mathfrak{A} = C_p, C_\infty, Q \text{ or } C_{p^\infty}$$

(or some other admissible group), we obtain a number of striking groups. For example, we have:

Corollary 10. *Given a prime* $p >> 1$, *there exists an infinite group whose proper subgroups are all conjugate and have order* p.

Corollary 11. *There exists a 2-generator group* G *whose maximal subgroups are all conjugate and are isomorphic to the additive group of* $\mathbf{Q}$, *and if* H *is a maximal subgroup of* G, $g \notin H$, *then*

$$gHg^{-1} \cap H = 1.$$

It may be expected that graded conjugacy relations will also be useful in proving our conjecture that in the Burnside group variety $\mathfrak{B}_n$ of odd exponent $n >> 1$, the analogue of Higman's embedding theorem is valid, i.e. recursive groups of exponent n are embeddable in finitely presented (in $\mathfrak{B}_n$) groups of exponent n.

9. Operations on groups

9.1. *An operation* $\circ$ on the class of all groups is defined as a map that takes any set $\{G_\alpha \mid \alpha \in I\}$ of groups to a group G° which contains the G_α, $\alpha \in I$, as subgroups and is generated by them. G° is called the *o-product* of the G_α, $\alpha \in I$, and is also denoted by $\Pi^\circ G_\alpha$ (or $G_1 \circ \ldots \circ G_n$ if $|I| < \infty$).

The most important examples of operations on groups are the free and direct products of groups. These operations possess the four main properties formulated below for an operation $\circ$.

(1°) *Regularity.* If $N = \langle G_{\alpha'} \rangle^{G^\circ}$ is the normal closure of $G_{\alpha'}$ in G°, then

$$N \cap \langle G_\alpha \mid \alpha \neq \alpha' \rangle = 1$$

in G°.

(2°) *Functoriality.* Any system of homomorphisms

$$\phi_\alpha : G_\alpha \to H_\alpha,\ \alpha \in I,$$

can be extended to a homomorphism $\Phi : G^\circ \to H^\circ$.

(3°) *Heredity* (or Mal'tsev's property). Any system of embeddings

$$\psi_\alpha : H_\alpha \to G_\alpha$$

can be extended to an embedding $\Psi : H^\circ \to G^\circ$.

(4°) *Associativity*. Let $I = \bigcup_{\beta \in J} I_\beta$, where $I_\beta \bigcap_{\beta \neq \beta'} I_{\beta'} = \emptyset$. Then the group

$$\prod_{\beta \in J}^{\circ} \left(\prod_{\alpha \in I_\beta}^{\circ} G_\alpha \right)$$

is naturally isomorphic to $G^\circ = \prod_{\alpha \in I}^{\circ} G_\alpha$.

The most natural of the known operations on groups besides free and direct products are the verbal products of S Moran [51] and the so-called Gruenberg-Shmel'kin multiplications [52, 53]. Let us recall their definitions.

Let $F = \Pi^* G_\alpha$ be the free product of groups G_α, $\alpha \in I$, let $C(F)$ be *the Cartesian subgroup* of F (i.e. the kernel of the natural mapping of F onto the direct product $\Pi^* G_\alpha$), and let $V(G)$ be the V-verbal subgroup of a group G for some set V of group words.

The *verbal V-product* $\Pi^V G_\alpha$ of groups G_α, $\alpha \in I$, is defined as the quotient group

$$\Pi^V G_\alpha = F/(V(F) \cap C(F)). \tag{9.1}$$

The Gruenberg-Shmel'kin (GS) V-multiplication $\Pi^{Ⓥ} G_\alpha$ of groups G_α, $\alpha \in I$, is defined as the quotient group

$$\Pi^{Ⓥ} G_\alpha = F/V(C(F)). \tag{9.2}$$

It is easy to see that both these operations are regular and functorial. S Moran [51] proved that the verbal V-product is associative and A L Shmel'kin [53] noted that it is not hereditary (at least when it differs from free and direct products). A L Shmel'kin [53-55] also proved that GS V-multiplication is a hereditary operation. On the other hand, S A Ashmanov and O N Matsedonska [56] have shown that GS V-multiplication is not an associative operation (when it differs from free and direct products).

9.2. The above-mentioned properties of verbal products and GS multiplications give a good illustration of the following old problem due to A I Mal'tsev (see [57]) which, however, preceded the appearance of these operations.

Do there exist regular hereditary and associative operations on the class of all groups different from free and direct products?

The weaker question (see [58, 59]) of the existence of non-trivial hereditary and associative operations on groups has also been studied.

The last question has been considered by S I Adyan. Extending the method of the classification of periodic words created by P S Novikov and S I Adyan [9, 10] on free products, S I Adyan [60, 61] introduced the so-called periodic n-products $\Pi^n G_\alpha$ of groups, for odd $n \geq 665$. However, these n-products can be defined only on the class of groups without involutions [61].

The *periodic n-product* $\Pi^n G_\alpha$ of groups G_α, $\alpha \in I$, without involutions is defined by S I Adyan as the quotient group F/P_n, where $F = \Pi^* G_\alpha$, and

$$P_n = \langle M_n \rangle^F$$

is the normal closure in F of the set M_n consisting of relators of the form f^n, $f \in F$, which are defined by a complicated induction similar to that in [9, 10]. S I Adyan [60] established that the periodic n-product

$$\Pi^n G_\alpha = F/P_n$$

possesses the following properties P1 and P2 (P0 holds by definition):

P0. $P_n = \langle f^n \mid \text{for certain words } f \in F \rangle^F$;

P1. $P_n \cap G_\alpha = 1$ for any $\alpha \in I$ (i.e. Π^n is indeed an operation);

P2. for each $g \in F/P_n$, either $g^n = 1$ or g is conjugate to an element of G_α for some α.

Moreover, S I Adyan [60] proved that the periodic n-product is a hereditary and associative operation, and in [61] showed that the group F/P_n is simple if the number of non-trivial factors G_α is more than one and $G_\alpha^n = G_\alpha$ for each $\alpha \in I$. Hence the n-products are not regular or functorial operations.

The second author [62] has defined an analogue of Adyan's n-product on the class of all groups (and for any $n >> 1$) by means of the following essentially simpler and shorter inductive construction.

Let $P_n(0) = 1$. Let us suppose that the subgroup

$$P_n(i) \subset F = \Pi^* G_\alpha$$

has been constructed, and let $F(i) = F/P_n(i)$, by definition. Then $S(i+1) \subset F$ is defined inductively as a maximal set (possibly empty) of words of length i+1 such that

(1) if a and b are two distinct elements of $S(i+1)$, then a is not conjugate to b or b^{-1} in $F(i)$;

(2) if $a \in S(i+1)$, then a is not conjugate to c^k in $F(i)$, where $|c| < \max(|a|, 2)$, $k \in \mathbf{N}$, and a is not equal to the product of two involutions in $F(i)$.

If $M_n(i+1) = \{a^n \mid a \in S(i+1)\}$, then

$$P_n(i+1) = \langle P_n(i) \cup M_n(i+1) \rangle^F,$$

and the normal subgroup $P_n(i+1)$ has been constructed for every $i \geq 0$.

Then the *periodic* $\bar{n}$*-product* $\bar{\Pi}^n G_\alpha$ of groups G_α, $\alpha \in I$, is the quotient group $F/\bar{P}_n$, where $\bar{P}_n = \bigcup_{i=0}^{\infty} P_n(i)$.

The analogy between Adyan's periodic n-products and our $\bar{n}$-products is that the group $F/\bar{P}_n$ possesses the following properties $\overline{P0}$ - $\overline{P2}$ resembling P0 - P2:

$\overline{P0}$. $\bar{P}_n = \langle f^n \mid$ for certain words $f \in F$,

where f and f^{-1} are not conjugate in $F/\bar{P}_n \rangle^F$;

$\overline{P1}$. $\bar{P}_n \cap G_\alpha = 1$ for any $\alpha \in I$;

$\overline{P2}$. for each $g \in F/\bar{P}_n$, either $g^n = 1$, or g is conjugate in $F/\bar{P}_n$ to either an element of G_α for some α, or to g^{-1} (the last possibility can be excluded when F has no involutions).

The question on hereditary associative operations on groups is answered affirmatively by:

Theorem 23 (A Yu Ol'shanskii [62]). *Given* n >> 1 (e.g. n > 10^{10}), *the periodic* $\bar{n}$*-product* $\bar{\Pi}^n$, *where*

$$\bar{\Pi}^n G_\alpha = F/\bar{P}_n,$$

is a hereditary, associative operation on the class of all groups and it differs from the free and direct products.

9.3. Despite the superficial similarity of periodic n- and $\bar{n}$-products, a question due to the second author on the coincidence of these products has turned out to be difficult. However, the answer is affirmative and in fact much more has been proved.

Let P be a subgroup of the free product $F = \Pi^* G_\alpha$ and let a subgroup $P \subset F$ possess the properties P0-P2 (in the formulation of which P_n is replaced by P). Then we shall call P briefly an n-*subgroup* of F. Similarly, if a subgroup $\bar{P} \subset F$ possesses the properties $\overline{P0}$ - $\overline{P2}$, then we shall call $\bar{P}$ an $\bar{n}$-*subgroup* of F.

Theorem 24 (S V Ivanov [68]).

(1) *Let* F *be a free product without involutions and* n *an odd number* >> 1 (e.g. n > 10^{10}). *Then there exists a unique* n-*subgroup of* F.

(2) *Let* F *be a free product and* n >> 1 (*e.g.* n > 10^{10}). *Then there exists a unqiue* $\bar{n}$-*subgroup of* F.

Corollary 12. *For odd* n >> 1, *the periodic* $\bar{n}$*-product coincides with the* n-*product on the class of groups without involutions.*

The non-triviality of Theorem 24 can be illustrated by the following two theorems, of which the first generalizes the above mentioned result of S I Adyan on the simplicity of n-products.

Theorem 25 (S V Ivanov [68]). *Let the free product* F *be non-degenerate (i.e. the number of non-trivial factors* G_α *of* F *is more than one and* F *is not a dihedral group), and let* n >> 1 (*e.g.* n > 10^{10}). *Then*

(1) *if* n *is odd and* F *has no involutions, then the quotient group* F^n/P_n *is simple, where*

$$F^n = \langle f^n \mid f \in F \rangle$$

and P_n *is an* n-*subgroup of* F:

(2) *the quotient group* $F^n/\bar{P}_n$ *is simple, where* $\bar{P}_n$ *is an* $\bar{n}$-*subgroup of* F.

Theorem 26 (S V Ivanov [68]). *Let* F *be a non-degenerate free product,* $n >> 1$ (e.g. $n > 10^{10}$). *Then*

(1) *if* n *is odd and* F *has no involutions, then there exists a continuous set of subgroups* $P \subset F$ *such that every* P *possesses the properties* P0, P1, $P \subset F^n$, *and the quotient group* F^n/P *is simple:*

(2) *there exists a continuous set of subgroups* $\bar{P} \subset F$ *such that every* $\bar{P}$ *possesses the properties* $\overline{P0}$, $\overline{P1}$, $\bar{P} \subset F^n$, *and the quotient group* $F^n/\bar{P}$ *is simple.*

9.4. The simplicity of n- and $\bar{n}$-products is, in fact, the degeneration of regularity and functoriality. Thus, for natural operations such as verbal products and GS multiplications, heredity and associativity are not combined, and the known non-trivial hereditary and associative operations are rather unnatural.

It was hard to imagine that a non-trivial operation on groups with properties 1°-4° (or 1°, 3°, 4° as in the Mal'tsev problem) could have an unnatural and complicated definition. Apparently for this reason, S Moran [63] conjectured that there were no such operations. The absence of any progress in this area for the last twenty years was another confirmation of Moran's conjecture. Nevertheless, some natural new operations called strictly verbal products have recently been found by the first author.

We begin with a definition of independent interest.

Let V be a set of words, let G be a group, and let H be a subgroup of G. Let us denote by $\tilde{V}(H)$ the subgroup of H generated by all values of words from V in the group G which belong to H, i.e.

$$\tilde{V}(H) = < v(g_1, ..., g_m) \mid v \in V, g_1, ..., g_m \in G, v(g_1, ..., g_m) \in H >. \quad (9.3)$$

It is clear that if H is a normal, characteristic, or fully invariant subgroup of G, then so is $\tilde{V}(H)$, as well as V(H).

The following inclusions are also obvious:

$$V(H) \subset \tilde{V}(H) \subset V(G) \cap H. \quad (9.4)$$

The *strictly verbal V-product* $\Pi^{\tilde{V}} G_\alpha$ of groups G_α, $\alpha \in I$, is defined to be a quotient group

$$\Pi^{\tilde{V}} G_\alpha = \mathbf{F} / V(C(\mathbf{F})), \tag{9.5}$$

where, as above, $\mathbf{F} = \Pi^* G_\alpha$ and $C(\mathbf{F})$ is the Cartesian subgroup of $\mathbf{F}$.

The inclusions (9.4), together with definitions (9.1), (9.2), (9.5), mean that there exist natural epimorphisms

$$\Pi^{\textcircled{V}} G_\alpha \rightarrow \Pi^{\tilde{V}} G_\alpha \rightarrow \Pi^{V} G_\alpha.$$

It is not hard to prove the following:

Proposition (S V Ivanov [64]). *For any set* V *of words, the strictly verbal* V-*product is a regular, functorial and associative operation.*

We note that if $V(G_\alpha) = 1$ for any $\alpha \in I$, then

$$\Pi^{V} G_\alpha = \Pi^{\tilde{V}} G_\alpha.$$

Hence, if one regards the verbal V-product as an extension of the free, $\mathcal{V}$-product in the group variety $\mathcal{V}$, where $\mathcal{V}$ is given by the laws $v = 1$, $v \in V$, to the class of all groups, then the strictly verbal V-product can be regarded in the analogous way.

Let $V \subset V'$ and $V(F_\infty) = V'(F_\infty)$ (where $V, V' \subset F_\infty$ and where F_∞ is the free group of countable rank), i.e. V and V' define the same variety. Then, obviously, for verbal products,

$$\Pi^{V} G_\alpha = \Pi^{V'} G_\alpha,$$

but for strictly verbal products one can only assert the existence of a natural epimorphism

$$\Pi^{\tilde{V}} G_\alpha \rightarrow \Pi^{\tilde{V}'} G_\alpha.$$

Moreover, if V' is a maximal subgroup with the property $V(F_\infty) = V'(F_\infty)$, then V' is verbal and

$$\Pi^{V'} G_\alpha = \Pi^{\tilde{V}'} G_\alpha.$$

Thus, the most interesting and free extension of the free $\mathcal{V}$-product would be when $\{v = 1, v \in V\}$ is some minimal basis for the laws of $\mathcal{V}$.

Let us consider strictly verbal x^n-products which we shall call briefly $\tilde{n}$-*products* and denote by $\Pi^{\tilde{n}}$.

If any group G_α, $\alpha \in I$, contains no non-trivial element g with $g^n = 1$, then it is easy to see that the $\tilde{n}$-product $\Pi^{\tilde{n}} G_\alpha$ coincides with the GS x^n-multiplication $\Pi^{\otimes} G_\alpha$. Hence in view of the heredity of GS multiplications we have the heredity of the $\tilde{n}$-product on the class of groups without non-trivial elements g of order dividing n.

With the help of the construction of a graded presentation for the group $\Pi^{\tilde{n}} G_\alpha$, the following theorem on the partial heredity of the $\tilde{n}$-product has been proved.

Theorem 27 (S V Ivanov [64, 1]). *For any odd* $n >> 1$ (*e.g.* $n > 10^{10}$), *the strictly verbal* x^n *-product is a hereditary operation on the class of groups without involutions.*

So the heredity of the $\tilde{n}$-products is a very natural but unproved fact. At the same time, to solve the problem of A I Mal'tsev, some non-trivial hereditary strictly verbal products have been constructed as follows.

Let n be a composite number, let d be a divisor of n such that $n^{1/2} \le d < n$, and let $m = [n^{1/2}] + 1$, where [ ] is the integer-part function.

Let us define the words $v_n(x,y)$, $w_n(x,y)$ by the following formulae:

$$v_n(x,y) = [[[x^{2d}, y^{2d}]^d, x]^d, y],$$

$$w_n(x,y) = [x^{2d}, y^{2d}]v_n^m [x^{2d}, y^{2d}]v_n^{m+1} \ldots$$

$$\ldots [x^{2d}, y^{2d}]v_n^{m+5} [x^{2d}, y^{2d}]^{-6} v_n^{m+6}. \tag{9.6}$$

The affirmative solution of the Mal'tsev problem is given by:

Theorem 28 (S V Ivanov [64]). *For any composite odd number* $n >> 1$ (*e.g.* $n > 10^{216}$), *the strictly verbal* $w_n(x,y)^n$ *-product (see (9.6)) is a regular, functorial, hereditary, associative operation on the class of all groups, which differs from free and direct products.*

The proof of Theorem 28 (of heredity, in fact) is based on the construction of a graded presentation for the strictly verbal $w_n(x,y)^n$ -product

$$G = \tilde{\Pi}^{w_n^{n}} G_\alpha$$

and on the following two properties of the word $w_n(x,y)$.

W1 If $f,g \in F = \Pi^* G_\alpha$ (or $f,g \in G$) and $w_n(f,g)$ is equal to the product of two involutions in **F** (in **G**), then $w_n(f,g)$ is conjugate to an element of some factor G_α of **F** (of **G**).

W2 Let $H_\alpha \subset G_\alpha$, $\alpha \in I$, be subgroups of the factors G_α of **F** (of **G**), let

$$K = \Pi^* H_\alpha \subset F \qquad (H = \langle H_\alpha \mid \alpha \in I \rangle \subset G)$$

and let some value v of the word $w_n(x,y)$ in the group **F** (in **G**) belong to the subgroup **K** (to **H**). Then if v is not a value of $w_n(x,y)$ in the subgroup **K** (in **H**), then v is conjugate to an element of some factor G_α of **F** (of **G**).

It appears to be very difficult to find a non-trivial word w with properties W1, W2 (even for the group **F** alone), and also to prove them for the group **G**. We note that the word (9.6), is the simplest known example of a word with properties W1, W2.

9.5. Just like the traditional V-extensions F/V(R) of F/R, where F is a free group, R is normal in F, and V is some set of words, the $\tilde{V}$-*extensions* F/$\tilde{V}$(R) of F/R (see definition (9.3)) are of much interest.

We formulate here only two algorithmic results on the groups F/R^n, $F/R^{\tilde{n}}$, where

$$R^n = x^n(R), \quad R^{\tilde{n}} = x^{\tilde{n}}(R).$$

Theorem 29 (S V Ivanov). *Let a group* F/R *contain no involution and let* n *be odd,* $n > 10^{10}$. *Then*

(1) *The following three conditions are equivalent:*

  (i) *the word problem is soluble in* F/R,

  (ii) *the word problem is soluble in* F/R^n,

  (iii) *the conjugacy problem is soluble in* F/R^n;

(2) *the following three conditions are equivalent:*

($\tilde{\text{i}}$) *the set of solutions of the equation* $x^n = 1$ *is recursive in* F/R,

($\tilde{\text{ii}}$) *the word problem is soluble in* $F/R^{\tilde{n}}$,

($\widetilde{\text{iii}}$) *the conjugacy problem is soluble in* $F/R^{\tilde{n}}$.

10. Factorizations

10.1. According to B H Neumann [65], a group G is called a general product of its subgroups A, B if

$$G = AB \text{ and } A \cap B = 1.$$

In this case we shall write $G = A \circ B$ ($\circ$ is not an operation as in Section 9).

We begin with the following two theorems.

Theorem 30 (S V Ivanov [66, 67]). *Let* G *be a finitely presented group satisfying the small cancellation condition* $C'(\lambda)$ *for* $\lambda = 10^{-3}$, *and let* H *be an arbitrary non-cyclic and non-dihedral subgroup of* G. *Then* H *is isomorphic to a general product* $F_\infty \circ F_\infty$, *where* F_∞ *is free of countable rank.*

Theorem 31 (S V Ivanov [66, 67]). *Let* G *be an arbitrary non-cyclic subgroup of the free countably-generated Burnside group* $\mathbf{B}(\infty,n)$ *of odd exponent* $n >> 1$ *(e.g.* $n > 10^{12}$*). Then* G *is isomorphic to a general product*

$$\mathbf{B}(\infty,n) \circ \mathbf{B}(\infty,n).$$

At first sight, it may seem strange that the proofs of Theorems 30, 31 are also based on an application of graded diagrams. But this is indeed the case, and the basis of the proofs of Theorems 30-34 is the construction of a graded presentation with an extended alphabet and with some set of defining relations which contains relations of a new "triangular" form AB = C, where A, B, C are specially chosen words with

$$|A| \approx |B| \text{ and } |A|, |B| >> |C|.$$

It should be pointed out that even the following corollary of Theorem 30 was hitherto unknown.

Corollary 13. *Let* G *be a finitely presented small cancellation group with condition* $C'(10^{-3})$ *and let* H *be an arbitrary non-cyclic non-dihedral subgroup of* G. *Then* H *contains a subgroup isomorphic to the free group* F_2.

The following known assertion has been obtained by V S Atabekyan [22] and has answered affirmatively a question raised by the second author (8.53b, [18]): does every non-cyclic subgroup of **B**(m,n) contain **B**(2,n)?

Corollary 14. *Every non-cyclic subgroup* H *of* **B**(m,n), *where* n *is odd,* n >> 1, *contains a subgroup isomorphic to* **B**(∞,n).

We also mention:

Corollary 15. *Any non-cyclic subgroup* H *of* F_∞ *is isomorphic to a general product* $F_\infty \circ F_\infty$.

Certainly, in view of the Nielsen-Schreier theorem, Corollary 15 in fact asserts that $F_m = F_\infty \circ F_\infty$, where $1 < m \leq \infty$. However, **B**(m,n) contains many nonfree subgroups (see Section 11.5). Therefore the similarity between Theorem 31 and Corollary 15 is worth emphasizing.

Similar descriptions also exist in a more general setting, for subgroups of free products and of free $\mathfrak{B}_n$-products (in the variety $\mathfrak{B}_n$ of groups of odd exponent n >> 1).

Theorem 32 (S V Ivanov [66, 67]). *Let* **F** *be a countable free product of groups* G_α, $\alpha \in I$, *and let* H *be an arbitrary non-cyclic and non-dihedral subgroup of* **F**. *Then* H *is either conjugate to a subgroup of some factor* G_α *or is isomorphic to a general product* $F_\infty \circ F_\infty$.

Theorem 33 (S V Ivanov [66, 67]). *Let* **B** *be a countable free* $\mathfrak{B}_n$*-product of groups* B_α, $\alpha \in I$, *of odd exponent* n >> 1, *and let* H *be a non-cyclic subgroup of* **B**. *Then* H *is either conjugate to a subgroup of some factor* B_α *or is isomorphic to a general product*

$$\mathbf{B}(\infty,n) \circ \mathbf{B}(\infty,n).$$

It turns out that subgroups of certain other infinite groups can also be described in terms of general products.

10.2. B Amberg [69] has raised the following two questions: does a product G = AB of two groups A, B satisfying the maximum (or minimum) condition for subgroups itself satisfy the maximum (or minimum) condition?

N S Chernikov (7.54, [18]) posed the question of the finiteness of rank of the product of two groups of finite rank.

Theorem 34, which is of independent interest, answers the questions above.

Theorem 34 (S V Ivanov [66]). *Let* G_1, G_2 *be groups of Theorem 1 of prime exponents* $p_1, p_2 >> 1$, *whose proper subgroups are all of orders* p_1, p_2 *respectively. Then every finite or countable group* H *can be embedded into some group* $G = G_1 \circ G_2$ *isomorphic to a general product of* G_1, G_2.

11. Other applications

11.1. J von Neumann [70] proved that every locally soluble group is amenable and a group containing a free non-cyclic subgroup is not amenable. The conjecture that any group without non-cyclic free subgroups is amenable, attributed to J von Neumann (see [71]), is disproved by:

Theorem 35 (A Yu Ol'shanskii [72]). *There exists a non-amenable 2-generator torsion (or torsion free) group* G *of unbounded exponent whose proper subgroups are all cyclic.*

In connection with the original Theorem 35, we mention a stronger result of S I Adyan [73] stating that the free Burnside group **B**(m,n), where n is odd and $n \geq 665$, and $m > 1$, is not amenable.

11.2. Since the known examples of finitely generated infinite simple groups are 2-generated (or it is hard to say anything about the minimal numbers of generators), J Wiegold (6.44, [18]) has asked for the construction of a finitely generated infinite simple group requiring more than two generators. The problem is solved by:

Theorem 36 (V S Guba [74]). *There exists a finitely generated simple group* G *in which all 2-generator subgroups are free.*

We note that the minimal number of generators of G is unknown. So the following question seems to be of interest. Do there exist infinite simple groups that are (m+1)-generated but not m-generated, for $m \geq 2$?

11.3. Let us consider the group E(m,n) defined as the quotient group $F_m/K(m,n)$, where

$$F_m = < a_1, a_2, \dots >$$

is a free group of finite or countable rank m, and

$$K(m,n) = < f^n \mid f \notin F'_m F^n_m >,$$

where F'_m is the commutator subgroup of F_m and

$F^n_m = < f^n \mid f \in F_m >$.

Theorem 37 (S V Ivanov, Yu S Semenov [75], A Yu Ol'shanskii [1]). *Let* n *be an odd number,* $n >> 1$ *(e.g.* $n > 10^{10}$*). Then the group* E(m,n) *contains a subgroup isomorphic to the free group of rank two, for example*

$$< [a,b], [a^2,b^2] > \subset E(m,n).$$

Moreover, if n *is prime, then the commutator subgroup* E'(m,n) *of* E(m,n) *is torsion-free, while every element* $e \notin E'(m,n)$ *has order* n.

This theorem is also of independent interest in view of the following two corollaries.

Corollary 16. *The 2-generator group* E(2,n) *for odd* $n >> 1$, *generates the variety of all groups, but every free group* F_m, $m < \infty$, *is not residually an* E(2,n)-*group.*

Corollary 17. *The non-trivial characteristic subgroup* $K(\infty,n)$ *of* F_∞ *contains no non-trivial verbal subgroups of* F_∞.

Corollary 17 is obvious. Corollary 16 follows from the easy fact that

$$[x_1^n, ..., x_m^n] \in \operatorname{Ker} \alpha$$

for every epimorphism

$$\alpha : F_m = < x_1, ..., x_m > \rightarrow E(2,n).$$

Corollary 16 disproves a conjecture of S Meskin [76] that the non-cyclic free group F_m is residually a $\mathfrak{C}$-group if $\mathfrak{C}$ is a class of 2-generated groups that generates the variety of all groups (obviously, F_∞ is residually a $\mathfrak{C}$-group).

Corollary 17 gives the first example of a non-trivial characteristic subgroup H of F_∞ such that H contains no non-trivial verbal subgroup of F_∞. The known examples of such subgroups (due to the second author [76], R M Bryant [77] and Yu G Kleiman [19, 26]) contain non-trivial verbal subgroups. (These examples disprove the conjecture of B H Neumann [78] that every characteristic subgroup of F_∞ is fully invariant, i.e. a verbal subgroup of F_∞.)

11.4. It is easy to prove that the Noethericity of the integral group algebra **Z**(G) of a group G implies the maximum condition for subgroups of G, i.e. that G is a max-group. On the other hand, P Hall's theorem [79] asserts that **Z**(G) is Noetherian if G is a polycyclic-by-finite group. Taking into account that before the paper [80] of the second author, all known max-groups were polycyclic-by-finite, the following question seems natural. Does there exist a max-group G such that **Z**(G) is not Noetherian? The question is attributed to P Hall [81] and is answered affirmatively by:

Theorem 38 (S V Ivanov [82, 1]). *There exists a 2-generated max-group* G *whose maximal subgroups are all cyclic but its integral group algebra is not Noetherian.*

Theorem 38 follows easily from the asphericity of the groups of Theorem 1, Theorem 10, and W Magnus' theorem on the embeddability of the relation module of G into a free **Z**(G)-module. In fact a strong result is obtained in [82]: there exists a max-group G* such that all the maximal subgroups of G* are cyclic and **Z**(G*) contains a right ideal which is a direct sum of countably many non-zero right ideals.

In connection with Theorem 38, we mention the very interesting known question on the existence of a non-polycyclic-by-finite group G such that **Z**(G) is Noetherian.

11.5. P S Novikov and S I Adyan (see [10]) proved not only the infiniteness of the free Burnside groups **B**(m,n), for $m > 1$, and odd exponent $n \geq 665$, but also obtained much important information on the groups **B**(m,n) (see also the papers [21, 73] of S I Adyan). Here we would like to draw attention to some new results about the groups **B**(m,n) (see also Theorem 31).

Theorem 39 (V S Atabekyan [22]). *Let* n *be odd,* $n >> 1, 2 \leq k \leq m$. *Then the group* **B**(m,n) *is*

(1) *residually a* **B**(k,n)*-group,*

and

(2) *residually a C-group, where C is the class of all simple groups in which all proper subgroups are finite.*

In Theorem 40, the following definitions are used. Let G be a subgroup of **B**(m,n). We define the *(torsion) root* $\sqrt{G}$ of G in **B**(m,n) as the subgroup generated by all elements b such that for each b there exists a k, such that $b^k \in G$ and $b^k \neq 1$. We shall say that G is an *isolated* subgroup if $G = \sqrt{G}$, and G is a *free* subgroup if G is isomorphic to a free Burnside group of the same exponent n.

Theorem 40 (S V Ivanov [83]). *Let* n *be a composite odd number,* n >> 1, *and let the rank* m *be finite or countable. Then*

(1a) *if* n *is not a prime-power, i.e.* $n \neq p^t$, *then* $\sqrt{G}$ *equals the subgroup* $N^*(G)$ *generated by the normalizers in* **B**(m,n) *of all non-trivial subgroups of* G, *i.e.*

$$\sqrt{G} = N^*(G) := \langle N_{\mathbf{B}(m,n)}(H) \mid H \subset G, H \neq 1 \rangle;$$

(1b) *if* n *is a prime-power, then the following inclusions are valid*

$$\sqrt{G} \subset N^*(G) \subset \sqrt{\sqrt{G}};$$

(2) *if* G *is a non-trivial free subgroup or a retract of* **B**(m,n), *then* G *is isolated.*

Under the hypotheses of Theorem 40, we obviously have two corollaries.

Corollary 18. *A subgroup* $G \subset$ **B**(m,n) *is isolated if and only if the normalizer in* **B**(m,n) *of any subgroup* $H \neq 1$ *of* G *is contained in* G.

Corollary 19. *If* G *is a non-trivial free subgroup of* **B**(m,n), *then the normalizer in* **B**(m,n) *of any non-trivial subgroup of* G *is contained in* G.

We note that Theorem 40 was stimulated by the still open question (7.1 [18]) of S I Adyan on the non-freeness of proper normal subgroups of **B**(m,n), where n is prime, $n \geq 665$. (By Corollary 19, this is true for odd composite n >> 1.)

In Section 10.1, Corollary 14, we saw that **B**(m,n) contains many free subgroups. The following theorems show how free generators of free subgroups can be chosen.

Theorem 41 (S V Ivanov). *Let* b, c *be elements of the group* **B**(m,n), *where* n *is odd and* n >> 1, *let the subgroup* < b,c > *be non-cyclic, and let the words* $w_{V_k}(x,y)$ *be defined by formulae (5.4). Then the elements*

$$w_{V_k}(b,c),\ k \in \mathbf{N},$$

freely generate a free subgroup of **B**(m,n).

Theorem 42 (S V Ivanov). *Let* m > 1, *and* n *be odd and* n >> 1. *Then every element* b *of the group* **B**(m,n) *of order* n *can be included in some infinite set*

$$S \subset \mathbf{B}(m,n)$$

that freely generates a free subgroup of **B**(m,n).

It seems reasonable to stop here and make some concluding remarks. Firstly, it should be pointed out that the words "there exist" in the statements of many theorems in the paper could be replaced by the words "can be effectively constructed" and that the word problem and conjugacy problem are soluble for many of the groups considered above. We also note that the idea of graded diagrams appeared originally in the papers of the second author [80] and E Rips [43]. Finally, we apologise for not being able to give full references for each question, as this would make the paper too long.

Acknowledgements

The second author is very grateful to the organisers of the conference, Dr C M Campbell and Dr E F Robertson for their warm hospitality.

The authors would like to thank Dr V Shpil'rain, who read the manuscript and made some valuable remarks.

Note added in proof

The Editors would like to thank Dr D L Johnson for his editorial assistance with this paper.

References

1. A Yu Ol'shanskii, *Geometry of defining relations in groups* (Nauka, Moscow, 1989).
2. E R van Kampen, On some lemmas in the theory of groups, *Amer. J. Math.* **55** (1933), 268-273.
3. R C Lyndon & P E Schupp, *Combinatorial group theory* (Springer-Verlag, Heidelberg, 1977).
4. M Gromov, *Hyperbolic groups,* in *Essays in Group Theory* (New York, 1987), 75-263.
5. P E Schupp, On Dehn's algorithm and the conjugacy problem, *Math. Ann.* **178** (1968), 119-130.
6. A Yu Ol'shanskii, Diagrams of homomorphisms of surface groups, *Sibirsk. Mat. J.* **30** (1989), 150-171.
7. A Yu Ol'shanskii, Groups of bounded exponent with subgroups of prime order, *Algebra i Logika* **21** (1982), 553-618.
8. A Yu Ol'shanskii, On a geometric method in combinatorial group theory, *Proc. Int. Congress Math. I* (1983), 415-424.

9. P S Novikov & S I Adyan, On infinite periodic groups I, II, III, *Izvestiya Akad. Nauk SSSR, Ser. Mat.* **32** (1968), 212-244, 521-524, 709-731.

10. S I Adyan, *The Burnside problem and identities in groups* (Nauka, Moscow, 1975).

11. A Yu Ol'shanskii, On the Novikov-Adyan theorem, *Mat. Sbornik* **118** (1982), 203-235.

12. S I Adyan, Infinite irreducible systems of group laws, *Izvestiya Akad. Nauk SSSR, Ser. Mat.* **34** (1970), 715-734.

13. S V Ivanov, P Hall's conjecture on the finiteness of verbal subgroups, *Izvestiya Vuzov. Matem.* **6** (1989), 50-60.

14. H Neumann, *Varieties of groups* (Springer-Verlag, Heidelberg, 1967).

15. A Yu Ol'shanskii, Varieties in which all finite groups are abelian, *Mat. Sbornik* **126** (1985), 59-82.

16. S V Ivanov, *On several questions on the theory of groups varieties* (Proc. Intern. Algebraic Conf. : Group Theory, Novosibirsk, 1989), 49.

17. S V Ivanov, On some questions in the theory of group varieties, to appear.

18. *Kourovkaya tetrad'* (unsolved questions of group theory) (10th Ed., Novosibirsk, 1986).

19. Yu G Kleiman, On some questions in the theory of group varieties, *Izvestiya Akad. Nauk. SSSR, Ser. Mat.* **47** (1983), 37-74.

20. *Sverdlovskaya tetrad'* (unsolved questions of semigroup theory) (2nd Ed., Sverdlovsk, 1979).

21. S I Adyan, Normal subgroups of free periodic groups, *Izvestiya Akad. Nauk SSSR, Ser. Mat.* **45** (1981), 931-947.

22. V S Atabekyan, On simple and free periodic groups, *Vestnik Moskov. Univ. Ser. I*, **6** (1987), 76-78.

23. A M Storozhev, Varieties of finite exponents in which all finite groups are abelian, to appear.

24. G Baumslag, Wreath products and extensions, *Math. Z.* **81** (1963), 193-209.

25. S V Ivanov, Non-hopfian relatively free groups, to appear.

26. Yu G Kleiman, On laws in groups, *Trudy Moskov. Mat. Obshch.* **44** (1982), 62-108.

27. A M Storozhev, *On some class of verbal subgroups* (Proc. All Union Symp. Group Theory, Sverdlovsk, 1989), 108-109.

28. I S Ashmanov & A Yu Ol'shanskii, On abelian and central extensions of aspherical groups, *Izvestiya Vuzov. Matem.* **11** (1985), 48-60.

29. S I Adyan, On some torsionfree groups, *Izvestiya Akad. Nauk SSSR Ser. Mat.* **35** (1971), 459-468.

30. A Yu Ol'shanksii, A note on a countable non-topologised group, *Vestn. Mosk. Univ. Ser. I*, **3** (1980), 103.

31. A A Markov, On unconditionally closed sets, *Mat. Sbornik* **18** (1946), 3-28.

32. S Shelah, On a problem of Kurosh, the Jónsson group, and applications, in *Word Problems*, II (North-Holland, Amsterdam, 1980), 373-394.

33. R F Turner-Smith, Marginal subgroup properties for outer commutator words, *Proc. London Math. Soc.* **14** (1964), 321-341.

34. D J Collins & J Huebschmann, Spherical diagrams and identities among relations, *Math. Ann.* **261** (1982), 155-183.

35. S V Ivanov, Relation modules and relation bimodules of groups, semigroups and associative algebras, to appear.

36. S V Ivanov, The word problem for 1-relator semigroups is soluble, to appear.

37. V N Obraztsov, *A theorem on group embeddings and some corollaries* (Proc. XIX All Union Algebraic Conf. I, Lvov, 1987), 203.

38. S V Ivanov, *Geometric methods of studying groups with prescribed properties of subgroups* (Ph.D. Thesis, Moscow State University, 1988).

39. A Yu Ol'shanskii, Thrifty embeddings of countable groups, *Vestnik. Moskov. Univ. Ser. I*, **2** (1989), 28-34.

40. G Higman, B H Neumann & H Neumann, Embedding theorems for groups, *J. London Math. Soc.* **24** (1949), 247-254.

41. G S Deryabina & A Yu Ol'shanskii, Subgroups of quasi-finite groups, *Uspekhi Mat. Nauk* **41** (1986), 169-170.

42. A G Kurosh & S N Chernikov, Solvable and nilpotent groups, *Uspekhi Mat. Nauk* **2** (1947), 18-59.

43. E Rips, Generalized small cancellation theory and applications I : The word problem, *Israel J. Math.* **41** (1982), 1-146.

44. V Dlab & V Kořinek, The Frattini subgroups of a direct product of groups, *Czech. Math. J.* **10** (1960), 350-358.

45. V S Guba, Finitely generated divisible groups, *Izvestiya Akad. Nauk SSSR, Ser. Mat.* **50** (1986), 883-924.

46. S V Ivanov, *Three theorems on embeddings of groups* (Proc. All Union Symp. Group Theory, Sverdlovsk, 1989), 51-52.

47. S V Ivanov, Conjugacy relations and some theorems on embeddings of groups, to appear.

48. B H Neumann, Adjuction of elements to groups, *J. London Math. Soc.* **18** (1943), 4-11.

49. Yu N Gorchinskii, Groups with finitely many conjugacy classes, *Mat. Sbornik 31* (1952), 167-182.

50. P Hall, Some constructions for locally finite groups, *J. London Math. Soc.* **34** (1959), 305-319.

51. S Moran, Associative operations on groups I, *Proc. London Math. Soc.* **6** (1956), 581-596.

52. K W Gruenberg, Residual properties of infinite soluble groups, *Proc. London Math. Soc.* **7** (1957), 29-62.

53. A L Shmel'kin, On the theory of regular operations on groups, *Mat. Sbornik* **51** (1960), 277-292.

54. A L Shmel'kin, On solvable products of groups, *Sibirsk. Mat. J.* **6** (1965), 212-220.

55. A L Shmel'kin, On free products of groups, *Mat. Sbornik* **79** (1969), 616-620.

56. S A Ashmanov & O N Matsedonska, On regular operations satisfying Mal'tsev's postulate, *Sibirsk. Mat. J.* **7** (1966), 1216-1229.

57. O N Golovin, Metabelian products of groups, *Mat. Sbornik* **28** (1951), 431-444.

58. A G Kurosh, *Group Theory, 3rd Ed.* (Nauka, Moscow, 1967).

59. O N Golovin & M A Bronshtein, *Axiomatic classification of exact operations, Selected problems of algebra and logic* (Nauka, Novosibirsk, 1973), 40-96.

60. S I Adyan, Periodic products of groups, *Trudy Mat. Inst. Akad. Nauk SSSR* **142** (1976), 3-21.

61. S I Adyan, On the simplicity of periodic products of groups, *Dokl. Akad. Nauk SSSR* **241** (1978), 745-748.

62. A Yu Ol'shanskii, The problem of A I Mal'tsev on operations on groups, *Trudy. Sem. Petrovsk.***14** (1989), 231-255.

63. S Moran, On a question of Mal'tsev, *Bull. Acad. Polon. Ser. Mat., Phys., Astr.* **9** (1961), 853-855.

64. S V Ivanov, Strictly verbal products of groups and the problem of A I Mal'tsev on operations on groups, *Trudy Moskov. Mat. Obshch.* **54** (1991), to appear.

65. B H Neumann, Decomposition of groups, *J. London Math. Soc.* **10** (1935), 3-6.

66. S V Ivanov, *On general products of groups* (Proc. All Union Symp. Group Theory, Sverdlovsk, 1989), 147-148.

67. S V Ivanov, Factorizations of subgroups of small cancellation and free Burnside groups, to appear.

68. S V Ivanov, On periodic n-products of groups, to appear.

69. B Amberg, Artinian and noetherian factorized groups, *Rend. Sem. Mat. Univ. Padova* **55** (1976), 105-122.

70. J von Neumann, Zur allgemeinen Theorie des Masses, *Fund. Math.* **13** (1929), 143-151.

71. F Greenleaf, *Invariant means on topological groups and their applications* (Van Nostrand, Reinhold Company, New York, 1969).

72. A Yu Ol'shanskii, On the question of the existence of invariant means on groups, *Uspekhi Mat. Nauk* **35** (1980), 199-200.

73. S I Adyan, Random walks on free periodic groups, *Izvestiya Akad. Nauk SSSR Ser. Mat.* **46** (1982), 1139-1149.

74. V S Guba, A finitely generated simple group with free 2-generated subgroups, *Sibirsk. Mat. J.* **27** (1986), 50-67.

75. S V Ivanov & Yu S Semenov, *On some groups with free subgroups, some questions of contemporary algebra* (Moscow State University, 1990), to appear.

76. A Yu Ol'shanskii, On characteristic subgroups of free groups, *Uspekhi Mat. Nauk* **29** (1974), 179-180.

77. R M Bryant, *Characteristic subgroups of free groups* (Lecture Notes in Math. **372** 1974), 141-149.

78. B H Neumann, On characteristic subgroups of free groups, *Math. Z.* **94** (1966), 143-151.

79. P Hall, Finiteness conditions for soluble groups, *Proc. London Math. Soc.* **4** (1954), 419-436.

80. A Yu Ol'shanskii, An infinite simple noetherian group without torsion, *Izvestia Akad. Nauk SSSR Ser. Mat.* **43** (1979), 1328-1393.

81. *Dnestrovskaya tetrad'* (unsolved questions in the theory of rings and modules) (1st Ed., Kishinev, 1969).

82. S V Ivanov, On group rings of noetherian groups, *Mat. Zametki* **47** (1989), 61-66.

83. S V Ivanov, On subgroups of free Burnside groups of odd composite period, *Vestn. Mosk. Univ. Ser. I*, **2** (1989), 7-11.

84. I N Zyabrev & E A Reznichenko, *On laws in connected topological groups* (Proc. XVII All Union Algebraic Conf. I, Kishinev, 1985), 208.

RATIONAL GROWTH OF WREATH PRODUCTS

D L JOHNSON

University of Nottingham, Nottingham NG7 2RD

Abstract

It is shown that the standard wreath product $H \wr G$ has rational growth provided that H has rational growth and G is finite. The growth functions are calculated explicitly in a few simple cases.

0. Introduction

The growth series of a group G with respect to a finite generating set X is defined as follows. For each $n \geq 0$, let c_n denote the number of distinct elements of G that can be written as a product of n elements of $X^{\pm}$ and no fewer. Then we call the power series

$$\gamma(t) = \sum_{n \geq 0} c_n t^n$$

the *growth series of* G *with respect to* X. In what follows, we work with 'natural' sets of generators whose stipulation, for the sake of notational convenience, is usually confined to the context.

Following results of Milnor [12] relating the curvature of a differentiable manifold to the growth of its fundamental group, interest centred on connections between the rate of growth of the sequence $\{c_n\}$ and structural properties of G (see, for example, [16], [13] and [1]). This culminated in the celebrated theorem of Gromov [7] and its spectacular proof [4] using non-standard analysis.

The case when $\gamma(t)$ is a rational function is a particularly favourable one. For example, if $t = 1$ is not a zero of γ, then $\gamma(1)^{-1}$ is often equal to the Euler characteristic of G [15], but not always [14]. On the other hand, if $t = 1$ is a pole of γ, then (under reasonable assumptions) the order of that pole is the Gel'fand-Kirillov dimension of G [9,10]. Agian, if $\gamma(t)$ is rational, the growth of G is either polynomial or exponential, according as all its poles lie on the unit circle or not [10].

It follows that, for groups of intermediate growth (see [6], for example), the growth function cannot be rational. Furthermore, the growth of groups with insoluble word problem cannot even be algebraic [3]. Thirdly, the second discrete Heisenberg group, with generators a, b, c, d, z and defining relations

$$[a,b] = [c,d] = z, \quad [a,c] = [a,d] = [b,c] = [b,d] = e, \quad z \text{ central},$$

has recently been shown [8] to have irrational growth. This completes the list of groups known (to me) with irrational growth. Our hope in studying the growth of wreath products was explicitly to obtain a simple example of such a group, and this paper describes the blighting of that hope.

The method used to compute the growth series is crude but fairly constructive, and applicable in other situations [5]. The idea is to find a *normal form* for the elements of the group in question (Section 2) that is *minimal* (Section 3), in the sense that no word in this form can be written as a shorter word; then we merely count (Section 4) the minimal normal forms of each given length. For ease of reference, we first carry this out in the known case, see [2], of the free product (Section 1), and conclude with some explicit computations (Section 5).

1. Free products

Let G, H be groups having growth series $\gamma(t)$, $\eta(t)$ with respect to finite generating sets X, Y, respectively. Then the growth series $\phi(t)$ of G*H with respect to $X \cup Y$ is easily calculated as follows. First, express the elements of G*H in normal form

$$h_0 g_1 h_1 g_2 h_2 \ldots g_n h_n g_{n+1}, \qquad n \geq 0, \tag{1}$$

where

$$h_0 \in H,\ g_{n+1} \in G; \quad h_1,\ldots,h_n \in H\backslash\{e\}; \quad g_1,\ldots,g_n \in G\backslash\{e\}.$$

By the normal form theorem for free products [11], this form is minimal provided that every g_i, h_i in (1) is written as a word of minimal length in $X^{\pm}$, $Y^{\pm}$, respectively. Then the set of words (1) for a fixed n has the growth series

$$\eta(t)((\gamma(t) - 1)(\eta(t) - 1))^n \gamma(t).$$

Summing over all $n \geq 0$ gives the growth series for G*H:

$$\begin{aligned}\phi(t) &= \sum_{n\geq 0} \eta(t)((\gamma(\tau) - 1)(\eta(t) - 1))^n\gamma(\tau) \\ &= \frac{\eta(t)\gamma(t)}{1 - (\gamma(\tau) - 1)(\eta(t) - 1)} \\ &= \left(\frac{\eta(t) + \gamma(t) - \eta(t)\gamma(t)}{\eta(t)\gamma(t)}\right)^{-1} \\ &= (\eta(\tau)^{-1} + \gamma(t)^{-1} - 1)^{-1},\end{aligned}$$

which is rational provided $\eta(t)$ and $\gamma(t)$ are. This may be neatly expressed by saying that the "reciprocal of the growth series minus one" is additive over free products.

2. The normal form for wreath products

It is convenient to regard the standard wreath product $H \wr G$ as a split extension

$$1 \to B \to H \wr G \to G \to 1,$$

where the base group B is the direct product of the conjugates gHg^{-1}, $g \in G$. It follows that any element $w \in G$ can be written in the form

$$w = \Big(\prod_{s\in S} sh_s s^{-1}\Big) \cdot g, \tag{2}$$

where

S is a finite subset of G, $g \in G$, and

$$\left.\begin{array}{l} h: S \to H \\ s \mapsto h_s \end{array}\right\} \text{ is any mapping,}$$

and that the form (2) is unique up to the ordering of S chosen in forming the product Π. To ensure absolute uniqueness, we assign an ordering $(s_1,s_2,\ldots,s_n)$ to $S = \{s_1,s_2,\ldots,s_n\}$ in such a way that (to ensure minimality) the number

$$m = l(s_1) + \sum_{i=1}^{n-1} l(s_i^{-1} s_{i+1}) + l(s_n^{-1} g) \tag{3}$$

is minimal over the n! possible orderings. (Here, l denotes minimal length in G with respect to some specific generating set.) Note that this minimum m depends only on S and g, not on the function h. Writing h_i for h_{s_i} $(1 \le i \le n)$, we obtain from (2) the normal form

$$w = \left(\prod_{i=1}^{n} s_i h_i s_i^{-1}\right) \cdot \; g, \tag{4}$$

with subscripts increasing, that is, S is ordered in such a way that m in (3) is minimal.

3. Proof of minimality

We now fix the finite generating sets X, Y, $X \cup Y$ (with $X \cap Y = \emptyset$) for G, H, $H \wr G$, respectively, and proceed as follows: reduce any word w' in $X \cup Y$ to the form (4) by a finite sequence of operations that

(a) preserve the element of $H \wr G$ represented, and

(b) do not increase the length.

As X and Y are disjoint, we can express w' as an alternating product

$$w' = g_1 h_1 \ldots g_n h_n g_{n+1}, \tag{5}$$

of words in $X^{\pm}$, $Y^{\pm}$, respectively, where only g_1 and g_{n+1} may be trivial (cf. (1) above), and henceforth regard w' as an element of $H \wr G$.

We must first ensure that the partial products

$$s_i := g_1 g_2 \ldots g_i, \quad 1 \le i \le n,$$

represent distinct elements of G. If this is not so, then we can find i and j such that

$$s_i = s_j,\ 1 \le i < j \le n, \text{ but}$$

$$s_i \ne s_k,\ i < k < j.$$

This implies that the segment

$$s = g_{i+1}\, h_{i+1} \ldots g_{j-1}\, h_{j-1}\, g_j$$

in the right-hand side of (5) is equal in $H \wr G$ to a product of conjugates of $h_{i+1},\ldots,h_{j-1}$ by non-identity elements of G, and so commutes with h_j. We can thus collect h_j to the left of s (and the right of h_i). This clearly does not violate (a) or (b) above, and the H-length of w' is decreased from n to n-1. After a finite number of such operations, the s_i in (4) will all be distinct.

The word w′ of (5) now has the form (4), with $S = \{s_1,\ldots,s_n\}$ and

$$g = g_1\, g_2 \cdots g_{n+1}.$$

Since the s_i in (4) are all distinct, there can be no cancellation between the h_i in re-ordering the product of the $s_i h_i s_i^{-1}$, $1 \leq i \leq n$, and so the ordering that minimises the right-hand side of (5) also minimises $l(w')$. But just such an ordering is specified in (4). It follows that (4), subject to the minimality of (3), is indeed a minimal normal form.

4. The main theorem

Now let $\eta(t)$ be the growth series of H with respect to Y and $\omega_{S,g}(t)$ that of the set of elements of the form (4) of $H \wr G$ for a given pair $S \subseteq G$, $g \in G$. Then

$$\omega_{S,g}(t) = t^m(\eta(t) - 1)^n, \qquad (6)$$

where $m = m(S,g)$ is the minimum in (3) and $n = |S|$. Now if $\eta(t)$ is rational, so is the right-hand side of (6). If, in addition, G is finite, the growth series of $H \wr G$ is just the finite sum

$$\omega(t) = \sum_{S \subseteq G} \sum_{g \in G} \omega_{S,g}(t) = \sum_{S \subseteq G} \sum_{g \in G} t^m(\eta(t) - 1)^n, \qquad (7)$$

with m,n as above. This is again rational, and we have proved the following result.

Theorem. *Let* G,H *be groups generated by finite sets* X,Y. *Assume that* H *has rational growth with respect to* Y *and that* G *has finite order. Then the standard wreath product* $H \wr G$ *has rational growth with respect to* $X \cup Y$.

5. Some explicit computations

The formula (7) can be used to compute the growth functions of the groups $H \wr Z_n$ ($n = 2,3,4$) and $H \wr Z$ (for example, when $H \cong Z_2$ or Z). The calculations are elementary and we omit them; the results are as follows, where in each case

(i) H has growth series $\eta = \eta(t)$ with respect to a finite generating set Y,

(ii) G is a cyclic group generated by x,

(iii) $\omega(t)$ is the growth function of $H \wr G$ with respect to $Y \cup \{x\}$.

Example 1. The growth function of $H \wr Z_n$ is equal to

$$(1+t)\eta(1 + t(\eta-1)), \qquad n = 2, \quad (8)$$

$$\eta(1 + 2t + (\eta-1)(1+2t)2t + (\eta-1)^2(2+t)t^2), \qquad n = 3, \quad (9)$$

$$\eta(1+t)(1 + t + (\eta-1)(2+t)t(1+t) + (\eta-1)^2 2t^2(1+2t) + (\eta-1)^3 2t^3), \quad n=4. \quad (10)$$

There seems to be some kind of pattern emerging here.

Example 2. The growth function of $H \wr Z$ is given by

$$\omega(t) = (1+\alpha)\eta(1 + \alpha + 2(t/(1-t) + \beta)), \quad (11)$$

where

$$\alpha = \frac{(\eta-1)t^2}{1 - \eta t^2}, \quad \beta = \frac{1}{1-t}\left(\frac{(\eta-1)(1+t)t}{1 - \eta t} - \alpha\right). \quad (12)$$

In the case when H is Z_2 ($\eta = 1+t$) or Z ($\eta = (1+t)/(1-t)$), this gives

$$\frac{(1+t)^3(1-t)^2(1+t+t^2)}{(1-t^2-t^3)^2(1-t-t^2)}, \quad \frac{(1+t)^3(1-t)^3(1+t^2)}{(1-t-t^2-t^3)^2(1-2t-t^2)}, \quad (13)$$

respectively. The second of these agrees (modulo a misprint) with the result on the last page of [10].

References

1. H Bass, The degree of polynomial growth of finitely-generated nilpotent groups, *Proc. London Math. Soc.* (3) **25** (1972), 603-614.
2. N Billington, *The growth of stratified groups* (Ph.D. thesis, La Trobe 1986).
3. J W Cannon, *The growth of the closed surface groups and the compact hyperbolic groups* (Madison), preprint.
4. L van den Dries & A J Wilkie, Gromov's theorem on groups of polynomial growth and elementary logic, *J. Algebra* **89** (1984), 349-374.
5. M Edjvet & D L Johnson, The growth of certain amalgamated free products and HNN-extensions, *J. Austral. Math. Soc.*, submitted.
6. R I Grigorchuk, On the Milnor problem of group growth, *Dokl. Akad. Nauk. SSSR* **271** (1983), 30-33.
7. M Gromov, Groups of polynomial growth and expanding maps, *Publ. Math. IHES* **53** (1981), 53-78.
8. F Grünewald, rumour.

9. M Krause & T Lenagan, *Growth of algebras and Gel'fand-Kirillov dimension* (Pitman, London 1985).

10. M Lorenz, *Gel'fand-Kirillov dimension and Poincaré series* (de Kalb 1987), preprint.

11. R C Lyndon & P E Schupp, *Combinatorial group theory* (Springer-Verlag, Berlin 1977).

12. J Milnor, A note on curvature and fundamental group, *J. Differential Geometry* **2** (1968), 1-7.

13. J Milnor, Growth of finitely-generated solvable groups, *J. Differential Geometry* **2** (1968), 447-448.

14. Walter Parry, Counter-examples involving growth series and Euler characteristics, *Proc. Amer. Math. Soc.* **102** (1988), 49-51.

15. N Smythe, Growth functions and Euler series, *Invent. Math.* **77** (1984), 517-531.

16. J A Wolf, Growth of finitely-generated solvable groups and curvature of Riemannian manifolds, *J. Differential Geometry* **2** (1968), 421-446.

THE MODULAR GROUP AND GENERALIZED FAREY GRAPHS

G A JONES & D SINGERMAN

University of Southampton, Southampton SO9 5NH

K WICKS

University of Hull, Hull HU6 7RX

1. Introduction

The *modular group*

$$\Gamma = PSL(2,\mathbf{Z}) = SL(2,\mathbf{Z})/\{\pm I\}$$

is the quotient of the unimodular group SL(2,Z) by its centre $\{\pm I\}$. Thus the elements of Γ are the pairs of matrices

$$\pm\begin{pmatrix} a & b \\ c & d \end{pmatrix} \quad (a,b,c,d \in \mathbf{Z},\ ad - bc = 1); \tag{1.1}$$

we will omit the symbol $\pm$, and identify each matrix with its negative.

It is both traditional and useful to represent Γ as a group of Möbius transformations of the upper half-plane

$$\mathcal{U} = \{z \in \mathbf{C} \mid \operatorname{Im} z > 0\},$$

with the element (1.1) acting by

$$z \mapsto \frac{az + b}{cz + d}. \tag{1.2}$$

For example, using the fact that this action of Γ is discontinuous, one can show that Γ is isomorphic to a free product $C_2 * C_3$; more specifically, Γ is generated by the elements

$$X = \begin{pmatrix} 0 & 1 \\ -1 & 0 \end{pmatrix}, \quad Y = \begin{pmatrix} 0 & -1 \\ 1 & 1 \end{pmatrix}, \tag{1.3}$$

with defining relations

$$X^2 = Y^3 = 1 \tag{1.4}$$

(see, for example, [5, §6.8], [8, §1.4]).

However, one can also use (1.2) to define actions of Γ on other sets; perhaps the most natural choice is the *extended set of rationals*, that is, the rational projective line

$$\hat{\mathbf{Q}} = \mathbf{Q} \cup \{\infty\}$$

(where we adopt the usual conventions regarding ∞, so that (1.1) sends ∞ to a/c, with $a/c = \infty$ if $c = 0$). We will investigate this action in the spirit of the theory of permutation groups, and will show how some graphs arising from this action are closely related to the Farey graph and the Farey sequences, and to some interesting tessellations of surfaces.

The paper is organised as follows:

In Section 2 we show that Γ acts transitively but imprimitively on $\hat{\mathbf{Q}}$; in fact, we show that for each positive integer $N \neq 2,5$, there is a Γ-invariant equivalence relation on $\hat{\mathbf{Q}}$ with N blocks (equivalence classes).

In 1967 Sims [10] introduced the idea of the *suborbital graphs* of a permutation group G acting on a set Ω: these are graphs (possibly directed) with vertex-set Ω, on which G induces automorphisms. In Section 3 we summarise Sims' theory, and determine the suborbital graphs for Γ on $\hat{\mathbf{Q}}$, obtaining a graph $\mathcal{G}_{u,n}$ on which Γ acts, for each $n \in \mathbf{N}$ and each unit $u \in U_n$. The simplest of these, $\mathcal{G}_{1,1}$ (studied in Section 4), is the *Farey graph* $\mathcal{F}$, in which vertices r/s, x/y $\in \hat{\mathbf{Q}}$ are adjacent if and only if $ry - sx = \pm 1$. By representing the edges of $\mathcal{F}$ as hyperbolic geodesics in $\mathcal{U}$, we obtain a Γ-invariant triangulation $\mathcal{T} = \mathcal{T}_{1,1}$ of $\mathcal{U}$, isomorphic to the *Farey tessellation* of the unit disc introduced by Hurwitz [4] in connection with binary quadratic forms. More recently, $\mathcal{T}$ has been shown by the second author [11] to be a universal object for triangular maps on surfaces; it plays an important role in the work of Series [9] on ergodic properties of geodesics and continued fractions, and (via its dual $\mathcal{T}^*$) in the work of Biggs [1] on constructing graphs of large girth.

In Section 5 we investigate how the properties of $\mathcal{F}$ may be generalised to the other suborbital graphs. Each $\mathcal{G}_{u,n}$ is a disjoint union of $\psi(n) = n \prod_{p|n} \left(1 + \frac{1}{p}\right)$ isomorphic copies of a subgraph $\mathcal{F}_{u,n}$. Now $\mathcal{F}_{u,n}$ (which is isomorphic to a subgraph of $\mathcal{F}$) induces a tessellation $\mathcal{T}_{u,n}$ of $\mathcal{U}$ which is invariant under the congruence subgroup $\Gamma_0(n)$ of index $\psi(n)$ in Γ; for instance, $\mathcal{T}_{1,2}$ plays the same universal role for all maps as $\mathcal{T}$ does for triangular maps. The Farey graph $\mathcal{F} = \mathcal{F}_{1,1}$ is a connected, undirected graph with triangular circuits; the corresponding properties of the graphs $\mathcal{F}_{u,n}$ are more elusive, but we are able to show that $\mathcal{F}_{u,n}$ is connected if and only if $n \leq 4$, is undirected if and only if

$$u^2 + 1 \equiv 0 \bmod n,$$

and contains triangles if and only if $u^2 \pm u + 1 \equiv 0 \bmod n$.

Finally, a comment about presentation. Because this work involves a rather unusual confluence of several areas of mathematics (number theory, group theory, graph theory, hyperbolic geometry, for instance), we have taken the trouble to explain in some detail a few elementary ideas which may be familiar to many (but probably not all) of our readers.

2. The action of Γ on $\hat{\mathbb{Q}}$

Every element of $\hat{\mathbb{Q}}$ can be represented as a reduced fraction x/y, with $x,y \in \mathbf{Z}$ and $(x,y) = 1$; since $x/y = -x/-y$, this representation is not unique. We represent ∞ as $1/0 = -1/0$. The action (1.2) of Γ on $\hat{\mathbb{Q}}$ now becomes

$$\begin{pmatrix} a & b \\ c & d \end{pmatrix} : \ \frac{x}{y} \mapsto \frac{ax + by}{cx + dy} . \qquad (2.1)$$

Note that as

$$c(ax + by) - a(cx + dy) = -y$$

and

$$d(ax + by) - b(cx + dy) = x,$$

it follows that $(ax + by, cx + dy) = 1$ and so $(ax + by)/(cx + dy)$ is a reduced fraction. Of course, the actions of an element of Γ on x/y and on -x/-y are identical, as are the actions of a matrix and its negative, so we have a well-defined action of Γ on $\hat{\mathbb{Q}}$. The following result is elementary and well-known:

Lemma 2.1. (i) *The action of* Γ *on* $\hat{\mathbf{Q}}$ *is transitive.*

(ii) *The stabilizer of a point is an infinite cyclic group.*

Proof. (i) We shall prove that the orbit containing $\infty = 1/0$ is $\hat{\mathbf{Q}}$. If $x/y \in \hat{\mathbf{Q}}$ (in reduced form) then as $(x,y) = 1$ there exist $u,v \in \mathbf{Z}$ with $ux - vy = 1$. Then the element $\begin{pmatrix} x & v \\ y & u \end{pmatrix}$ of Γ sends ∞ to x/y.

(ii) By (i), the stabilizers of any two points in $\hat{\mathbf{Q}}$ are conjugate in Γ, so it is sufficient to consider the stabilizer Γ_∞ of ∞. This is easily seen to consist of the elements of the form $\begin{pmatrix} 1 & b \\ 0 & 1 \end{pmatrix}$, with $b \in \mathbf{Z}$, so Γ_∞ is the infinite cyclic subgroup generated by the element

$$Z = XY = \begin{pmatrix} 1 & 1 \\ 0 & 1 \end{pmatrix}.$$

We now consider the imprimitivity of the action of Γ on $\hat{\mathbf{Q}}$, beginning with a general discussion of primitivity of permutation groups. Let (G,Ω) be a permutation group, consisting of a group G acting on a set Ω. An equivalence relation $\approx$ on Ω is called G-*invariant* if, whenever $\alpha,\beta \in \Omega$ satisfy $\alpha \approx \beta$, then $g(\alpha) \approx g(\beta)$ for all $g \in G$; the equivalence classes are called *blocks*. Obvious examples of such relations are:

(i) the *identity relation*, $\alpha \approx \beta$ if and only if $\alpha = \beta$,

and

(ii) the *universal relation*, $\alpha \approx \beta$ for all $\alpha,\beta \in \Omega$.

We call (G,Ω) *imprimitive* if Ω admits some G-invariant equivalence relation other than (i) and (ii); otherwise, we call (G,Ω) *primitive*. Clearly, a primitive group must be transitive, for if not the orbits would form a system of blocks. The converse is false, but we have the following useful result.

Proposition 2.2 [2, Theorem 1.6.5]. *Let* (G,Ω) *be transitive. Then* (G,Ω) *is primitive if and only if* G_α, *the stabilizer of a point* $\alpha \in \Omega$, *is a maximal subgroup of* G *for each* $\alpha \in \Omega$.

Indeed, suppose that $G_\alpha < H < G$. Since G acts transitively, every element of Ω has the form $g(\alpha)$ for some $g \in G$. One easily checks that there is a well-defined G-invariant equivalence relation $\approx$ on Ω, given by

$$g(\alpha) \approx g'(\alpha) \text{ if and only if } g' \in gH.$$

If $\beta \in \Omega$, then $\beta = g(\alpha)$ for some $g \in G$, so the block $[\beta]$ containing β is $\{gh(\alpha) \mid h \in H\}$; in particular, the block $[\alpha]$ is the H-orbit

$$H(\alpha) = \{h(\alpha) \mid h \in H\}.$$

If $\{\ell_i \mid i \in I\}$ is a set of left coset representatives for H in G, then the blocks are the images $\ell_i H(\alpha)$ $(i \in I)$ of $H(\alpha)$; thus the number of blocks is equal to the index $|I| = |G:H|$ of H in G. There is an obvious induced action of G on the set $\Omega/\approx$ of blocks, the stabilizer of the block $[\alpha]$ being the subgroup H.

We now apply these ideas to the case where G is the modular group Γ, and Ω is $\hat{Q}$. Here Γ_∞, the stabilizer of ∞, is the subgroup of Γ generated by Z, so by finding subgroups H of Γ containing Γ_∞ (or equivalently, containing Z), we can produce Γ-invariant equivalence relations on $\hat{Q}$. Since $Z = \begin{pmatrix} 1 & 1 \\ 0 & 1 \end{pmatrix}$, some obvious choices for H are the congruence subgroups

$$\Gamma_0(n) = \left\{ \begin{pmatrix} a & b \\ c & d \end{pmatrix} \in \Gamma \mid c \equiv 0 \bmod n \right\},$$

where $n \in N$. Clearly $\Gamma_\infty < \Gamma_0(n) \leq \Gamma$ for each n, and the second of these inclusions is strict if $n > 1$, so Γ acts imprimitively on $\hat{Q}$. Let $\underset{n}{\approx}$ (or simply $\approx$) denote the Γ-invariant equivalence relation induced on $\hat{Q}$ by $\Gamma_0(n)$. If $v = r/s$ and $w = x/y$ are elements of $\hat{Q}$, then $v = g(\infty)$ and $w = g'(\infty)$ for elements $g, g' \in \Gamma$ of the form

$$g = \begin{pmatrix} r & * \\ s & * \end{pmatrix}, \quad g' = \begin{pmatrix} x & * \\ y & * \end{pmatrix};$$

now $v \approx w$ if and only if $g^{-1}g' \in H = \Gamma_0(n)$, and since

$$g^{-1} = \begin{pmatrix} * & * \\ -s & r \end{pmatrix}$$

we see that $v \approx w$ if and only if

$$ry - sx \equiv 0 \bmod n. \tag{2.2}$$

To put this another way, $v = r/s$ and $w = x/y$ are equivalent if and only if they "have the same reduction mod n", that is,

$$x \equiv ur \quad \text{and} \quad y \equiv us \bmod n \tag{2.3}$$

for some unit $u \in U_n$. (To see this, note that both (2.2) and (2.3) define Γ-invariant equivalence relations on $\hat{Q}$; the block containing ∞ is the same in each case, namely $[\infty] = \{x/y \in \hat{Q} \mid y \equiv 0 \bmod n\}$, so the two relations are equal.)

By our general discussion of imprimitivity, the number $\psi(n)$ of equivalence classes under $\underset{n}{\approx}$ is given by

$$\psi(n) = |\Gamma : \Gamma_0(n)|;$$

the following formula for $\psi(n)$ is well-known ([8, §4.3] for example), but for completeness we will sketch a proof here.

Lemma 2.3. $\psi(n) = n \prod_{p|n} \left(1 + \frac{1}{p}\right)$, *where the product is over the distinct primes* p *dividing* n.

Proof. First we show that ψ is a multiplicative function. Let $n = \ell m$ with $(\ell, m) = 1$. By (2.2),

$$v \underset{n}{\approx} w \text{ if and only if } v \underset{\ell}{\approx} w \text{ and } v \underset{m}{\approx} w,$$

so by counting equivalence classes we have

$$\psi(n) = \psi(\ell)\psi(m),$$

as required. Now the function

$$n \mapsto n \prod \left(1 + \frac{1}{p}\right)$$

on the right-hand side is clearly also multiplicative, so to prove the lemma it is sufficient to consider the case where n is a power of some prime p.

If $v = r/s \in \hat{Q}$, then either r or s is coprime to p, and is therefore a unit mod n; by (2.3), with u the inverse of this unit, we see that $v \approx 1/i$ or $j/1$ for some $i \in Z_n$ or $j \in Z_n$. It is easily checked that the 2n classes of $[1/i]$, $[j]$ are distinct, except that $[1/i] = [j]$ if and only if $ij \equiv 1 \bmod n$. The number of such coincident pairs is Euler's function

$$\phi(n) = |U_n| = n\left(1 - \frac{1}{p}\right),$$

so the number of distinct classes is

$$2n - \phi(n) = n\left(1 + \frac{1}{p}\right)$$

as required.

In particular, if n is a prime p, then there are $\psi(p) = p+1$ blocks, these being $[0],[1],\ldots,[p-1],[\infty]$, where

$$[j] = \{x/y \in \hat{Q} \mid x \equiv jy \bmod p\} \qquad (j \neq \infty),$$

$$[\infty] = \{x/y \in \hat{Q} \mid y \equiv 0 \bmod p\}.$$

The induced action of Γ on these blocks is the same as that of its quotient-group PSL(2,p) on the projective line $GF(p) \cup \{\infty\}$. For p = 2, the blocks have the form

$$[0] = \begin{Bmatrix}\text{even}\\ \text{odd}\end{Bmatrix}, \qquad [1] = \begin{Bmatrix}\text{odd}\\ \text{odd}\end{Bmatrix}, \qquad [\infty] = \begin{Bmatrix}\text{odd}\\ \text{even}\end{Bmatrix},$$

and Γ acts as $PSL(2,2) \cong S_3$ on them.

It follows from Lemma 2.3 that $\psi(n)$ is even if $n > 2$, while $\psi(1) = 1$ and $\psi(2) = 3$. However, by considering subgroups H other than $\Gamma_0(n)$ we can produce Γ-invariant equivalence relations on $\hat{Q}$ with an almost arbitrary number of blocks:

Theorem 2.4. *For each positive integer* $N \neq 2,5$ *there is a* Γ*-invariant equivalence relation on* $\hat{Q}$ *with* N *blocks.*

Proof. By our earlier comments on imprimitivity, it is sufficient to find a subgroup H of index N in Γ, containing the element $Z = XY$. Since Γ has the presentation

$$\Gamma = \mathrm{gp}\langle X,Y \mid X^2 = Y^3 = 1\rangle,$$

the conjugacy classes of subgroups H of index N in Γ correspond to transitive permutation groups of degree N having generators x,y (the permutations induced by X and Y) satisfying $x^2 = y^3 = 1$; then H (or some conjugate) will contain $Z = XY$ if and only if the permutation $z = xy$ has a fixed point in this representation.

The case N = 1 (corresponding to the universal equivalence relation) is trivial, so assume that N > 1. We will represent the required permutation group of degree N by a graph on N vertices (the points permuted), the nontrivial cycles of X being represented by undirected edges, and those of y by directed triangles (this graph is simply a coset diagram for H in Γ). If $N \equiv 0 \bmod 3$, say $N = 3k \geq 3$, then the graph

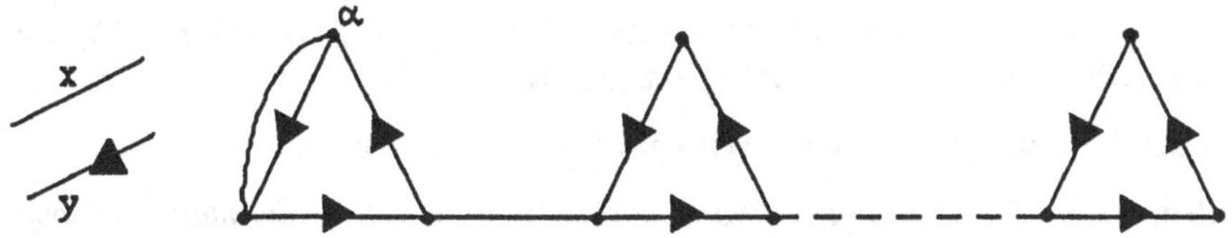

Fig. 1

in Fig. 1, with k directed triangles, represents a group of degree N generated by x (which has k-1 fixed points) and y (which has none), satisfying $x^2 = y^3 = 1$; since the graph is connected, this group is transitive; the vertex labelled α clearly satisfies $z(\alpha) = xy(\alpha) = \alpha$, so we can take $H = \Gamma_\alpha$. If $N = 3k + 1 \geq 4$, or if $N = 3k + 2 > 5$, then we simply modify the graph in Fig. 1 by adding one or two extra vertices (fixed by y) and joining each by an undirected edge to some vertex fixed by x.

This deals with all $N \neq 2,5$. The reader can easily verify that in each of the two remaining cases there is essentially just one connected graph one can draw, and that in the corresponding permutation group z has no fixed points. Thus the theorem cannot be extended to include these cases.

3. Suborbital graphs for Γ on $\hat{\mathbb{Q}}$

Let (G,Ω) be a transitive permutation group. Then G acts on Ω × Ω by

$$g : (\alpha,\beta) \mapsto (g(\alpha), g(\beta))$$

($g \in G$, $\alpha,\beta \in \Omega$). The orbits of this action are called *suborbitals* of G, that containing (α,β) being denoted by O(α,β). From O(α,β) we can form a *suborbital graph* $\mathcal{G}(\alpha,\beta)$: its vertices are the elements of Ω, and there is a directed edge from γ to δ if $(\gamma,\delta) \in O(\alpha,\beta)$.

Clearly $O(\beta,\alpha)$ is also a suborbital, and it is either equal to or disjoint from $O(\alpha,\beta)$. In the latter case, $\mathcal{G}(\beta,\alpha)$ is just $\mathcal{G}(\alpha,\beta)$ with the arrows reversed, and we call $\mathcal{G}(\alpha,\beta)$ and $\mathcal{G}(\beta,\alpha)$ *paired* suborbital graphs. In the former case, $\mathcal{G}(\beta,\alpha) = \mathcal{G}(\alpha,\beta)$ and the graph consists of pairs of oppositely directed edges; it is convenient to replace each such pair by a single undirected edge, so that we have an undirected graph which we call *self-paired.*

These ideas were first introduced by Sims [10], and are also described in a paper by Neumann [7] and in books by Tsuzuku [12] and by Biggs and White [2], the emphasis being on applications to *finite* groups.

The following general result is well-known and is easy to prove:

Proposition 3.1. *Let* $\mathcal{G}$ *be a suborbital graph for a transitive permutation group* (G,Ω). *Then*

(i) G *acts as a group of automorphisms of* $\mathcal{G}$;

(ii) G *acts transitively on the vertices of* $\mathcal{G}$;

(iii) *if* $\mathcal{G}$ *is self-paired, then* G *acts transitively on ordered pairs of adjacent vertices of* $\mathcal{G}$;

(iv) *if* $\mathcal{G}$ *is not self-paired, then* G *acts transitively on the edges of* $\mathcal{G}$.

Example. $O(\alpha,\alpha) = \{(\gamma,\gamma) \mid \gamma \in \Omega\}$ is the diagonal of $\Omega \times \Omega$. The corresponding suborbital graph $\mathcal{G}(\alpha,\alpha)$, called the *trivial* suborbital graph, is self-paired: it consists of a loop based at each vertex $\gamma \in \Omega$. We shall be mainly interested in the remaining *non-trivial* suborbital graphs.

We now investigate the suborbital graphs for the action of Γ on $\hat{\mathbb{Q}}$. Since Γ acts transitively on $\hat{\mathbb{Q}}$, each suborbital contains a pair (∞,v) for some $v \in \hat{\mathbb{Q}}$; writing $v = u/n$, with $n \geq 0$ and $(u,n) = 1$, we will denote this suborbital by $O_{u,n}$, and the corresponding suborbital graph $\mathcal{G}(\infty,v)$ by $\mathcal{G}_{u,n}$. If $v = \infty$, this is the trivial suborbital graph $\mathcal{G}_{1,0} = \mathcal{G}_{-1,0}$, so let us assume that $v \in \mathbb{Q}$. If $v' \in \mathbb{Q}$, then $O(\infty,v) = O(\infty,v')$ if and only if v and v' are in the same orbit of Γ_∞; since Γ_∞ is generated by

$$Z : v \mapsto v + 1,$$

this is equivalent to $v' = u'/n$ where $u \equiv u' \bmod n$. Thus $\mathcal{G}_{u,n} = \mathcal{G}_{u',n'}$ if and only if $n = n'$ and $u \equiv u' \bmod n$, so for each integer $n \geq 1$ there are $\phi(n)$ distinct suborbital graphs $\mathcal{G}_{u,n}$, one for each unit $u \in U_n$.

We will write $r/s \to x/y$ in $\mathcal{G}_{u,n}$ if $(r/s, x/y) \in O_{u,n}$, that is, if there is a directed edge from r/s to x/y in $\mathcal{G}_{u,n}$ (or an undirected edge if $\mathcal{G}_{u,n}$ is self-paired).

Theorem 3.2. $r/s \to x/y$ *in* $\mathcal{G}_{u,n}$ *if and only if either*

(a) $x \equiv ur \bmod n$, $y \equiv us \bmod n$, *and* $ry - sx = n$, *or*

(b) $x \equiv -ur \bmod n$, $y \equiv -us \bmod n$, *and* $ry - sx = -n$.

(If $sy > 0$, then (a) and (b) correspond to $r/s > x/y$ and $r/s < x/y$ respectively.)

Proof. If $r/s \to x/y$ in $\mathcal{G}_{u,n}$, then some element $\begin{pmatrix} a & c \\ b & d \end{pmatrix} \in \Gamma$ sends $\infty = 1/0$ to r/s, and u/n to x/y, so we have the matrix equation

$$\pm\begin{pmatrix} a & b \\ c & d \end{pmatrix}\begin{pmatrix} 1 & u \\ 0 & n \end{pmatrix} = \begin{pmatrix} r & x \\ s & y \end{pmatrix}. \tag{3.1}$$

If the plus sign is valid in (3.1), then $a = r$, $c = s$, $au + bn = x$ and $cu + dn = y$, so that $x \equiv ur$ and $y \equiv us \bmod n$; taking determinants in (3.1), we see that $ry - sx = n$, so (a) holds. Similarly, if the minus sign is valid, we obtain (b).

Conversely, if (a) holds then there exist integers b, d such that $x = ur + bn$ and $y = us + dn$, so that

$$\begin{pmatrix} r & b \\ s & d \end{pmatrix}\begin{pmatrix} 1 & u \\ 0 & n \end{pmatrix} = \begin{pmatrix} r & x \\ s & y \end{pmatrix}. \tag{3.2}$$

As $ry - sx = n$, we have $rd - bs = 1$, so $\begin{pmatrix} r & b \\ s & d \end{pmatrix} \in \Gamma$ and hence $r/s \to x/y$ in $\mathcal{G}_{u,n}$. If (b) holds, we have (3.2) with x and y replaced by $-x$ and $-y$, so that

$$r/s \to -x/-y = x/y$$

in $\mathcal{G}_{u,n}$.

Corollary 3.3. *The suborbital graph paired with* $\mathcal{G}_{u,n}$ *is* $\mathcal{G}_{-\bar{u},n}$ *where* $\bar{u}$ satisfies $u\bar{u} \equiv 1 \bmod n$.

Proof. Paired suborbital graphs have the same edges, with the arrows reversed. It is easy to verify that the conditions given in Theorem 3.2 for $r/s \to x/y$ in $\mathcal{G}_{u,n}$ are equivalent to those for $x/y \to r/s$ in $\mathcal{G}_{-\bar{u},n}$.

Corollary 3.4. $\mathcal{G}_{u,n}$ *is self-paired if and only if* $u^2 \equiv -1 \bmod n$.

In Section 2 we introduced, for each integer $n \geq 1$, a Γ-invariant equivalence relation $\underset{n}{\approx}$ on $\hat{\mathbb{Q}}$, with $r/s \underset{n}{\approx} x/y$ if and only if $ry - sx \equiv 0 \bmod n$. If $r/s \to x/y$ in $\mathcal{G}_{u,n}$, then Theorem 3.2 implies that $ry - sx = \pm n$, so $r/s \underset{n}{\approx} x/y$. Thus each connected component of $\mathcal{G}_{u,n}$ lies in a single block for $\approx_n$, of which there are $\psi(n)$, so we have:

Corollary 3.5. *$\mathcal{G}_{u,n}$ has at least $\psi(n)$ connected components; in particular, $\mathcal{G}_{u,n}$ is not connected if* $n > 1$.

In the next section, we will study the simplest non-trivial suborbital graph, namely $\mathcal{G}_{1,1}$, and show that it *is* connected.

4. The Farey graph

The graph $\mathcal{G}_{1,1}$ has $\hat{\mathbf{Q}}$ as its vertex set; by Corollary 3.4, it is self-paired, so we can regard it as an undirected graph. By Theorem 3.2, vertices r/s and x/y are adjacent if and only if $ry - sx = \pm 1$; for instance, the vertices adjacent to ∞ are the integers. By Proposition 3.1, Γ is a group of automorphisms of $\mathcal{G}_{1,1}$, acting transitively on vertices and on edges; in fact, we shall show that Aut $\mathcal{G}_{1,1}$ is the *extended modular group* PGL(2,Z), formed by allowing $ad - bc = \pm 1$ in (1.1).

We shall call $\mathcal{G}_{1,1}$ the *Farey graph*, and denote it by $\mathcal{F}$, because of its connection with Farey sequences. For each integer $m \geq 1$, the *Farey sequence* $\mathcal{F}_m$ of order m consists of all rational numbers x/y with $|y| \leq m$, arranged in increasing order. For example, $\mathcal{F}_4$ is:

$$\ldots, -\frac{1}{3}, -\frac{1}{4}, 0, \frac{1}{4}, \frac{1}{3}, \frac{1}{2}, \frac{2}{3}, \frac{3}{4}, 1, \frac{5}{4}, \frac{4}{3}, \ldots .$$

(Some authors restrict the elements of $\mathcal{F}_m$ to the intervals [0,1] or $[-m,m] \subset \mathbf{R}$; we shall not.) Clearly $\mathcal{F}_1 \subset \mathcal{F}_2 \subset \ldots$, and $\bigcup_{m \geq 1} \mathcal{F}_m = \mathbf{Q}$.

Lemma 4.1. *Let* r/s, x/y $\in \mathbf{Q}$ *be reduced rationals. Then the following three conditions are equivalent:*

(i) r/s *and* x/y *are adjacent vertices in* $\mathcal{F}$;

(ii) $ry - sx = \pm 1$;

(iii) r/s *and* x/y *are adjacent terms in* $\mathcal{F}_m$ *for some* m.

Proof. As we observed above, (i) and (ii) are equivalent by Theorem 3.2 (with $u = n = 1$). The equivalence of (ii) and (iii) is a standard result in elementary number theory [3, Ch.III], [6, Theorem 9.3], the relevant values of m being given by $\max(|s|, |y|) \leq m < |s| + |y|$.

This gives us a simple construction for $\mathfrak{F}$: we make each $\mathfrak{F}_m$ into a tree of valency 2 by joining each term to its immediate predecessor and successor; by the lemma, the union of these trees is the subgraph of $\mathfrak{F}$ induced on $\mathbf{Q}$, so we can form $\mathfrak{F}$ by adding a vertex labelled ∞ and joining it to the integers.

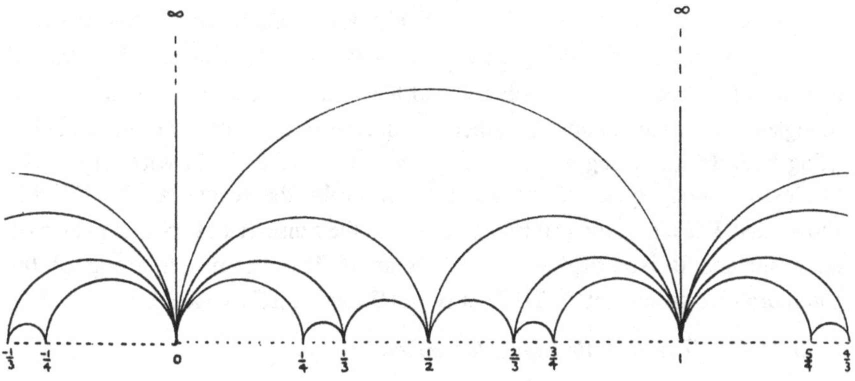

Fig. 2 : $\mathfrak{F} = \mathfrak{G}_{1,1}$

This construction is illustrated in Fig. 2, which shows those edges incident with ∞ or joining elements of $\mathfrak{F}_4$; the pattern is periodic, with period 1. For visual convenience (and for mathematical reasons which will become clear later), we have represented the edges of $\mathfrak{F}$ as hyperbolic geodesics in the upper half-plane

$$\mathcal{U} = \{z \in \mathbf{C} \mid \operatorname{Im} z > 0\},$$

that is, as euclidean semi-circles or half-lines perpendicular to $\mathbf{R}$. As usual, we regard half-lines as "meeting at ∞". We can use (1.2) to define an action by Γ as a group of hyperbolic isometries of $\mathcal{U}$; geodesics are sent to geodesics under this action, so our representation of $\mathfrak{F}$ in $\mathcal{U}$ is invariant under Γ. We now show that this representation is an *embedding*:

Corollary 4.2. *No edges of $\mathfrak{F}$ cross in $\mathcal{U}$.*

Proof. Suppose that two edges cross in $\mathcal{U}$. By Proposition 3.1 (iii), we may assume that one of them is the edge $\operatorname{Re} z = 0$ joining 0 and ∞, so the other must join rationals v and w where $v < 0 < w$. By Lemma 4.1, v and w are consecutive in some $\mathfrak{F}_m$, which is impossible since 0 will intervene.

We will call this embedding of $\mathcal{F}$ in $\mathcal{U}$ (or, more precisely, in $\mathcal{U} \cup \hat{\mathbb{Q}}$) the *Farey tessellation* $\mathcal{T}$ of $\mathcal{U}$. It is, in fact, a triangulation of $\mathcal{U}$ by hyperbolic triangles (with vertices in $\hat{\mathbb{Q}}$). Consider the region

$$\Delta = \{z \in \mathcal{U} \mid 0 < \mathrm{Re}\, z < 1, |z - \tfrac{1}{2}| > \tfrac{1}{2}\}$$

bounded by the edges joining 0, 1 and ∞. By Corollary 4.2, no other edges of $\mathcal{F}$ cross these three edges, so Δ is a face of $\mathcal{T}$ (a connected component of $\mathcal{U} \setminus \mathcal{F}$). Now Γ permutes the edges of $\mathcal{T}$ transitively, and contains an element (such as $z \mapsto -1/z$) which interchanges the two faces incident with an edge; thus Γ permutes the faces of $\mathcal{T}$ transitively, and hence they are all (like Δ) hyperbolic triangles. This triangulation (or rather, an equivalent triangulation of the unit disc, using Euclidean line-segments as edges) was introduced by Hurwitz [4] in order to classify binary quadratic forms; more recently, the second author [11] has shown that $\mathcal{T}$ is the "universal triangulation" in the sense that every triangular map on a surface is isomorphic to a quotient of $\mathcal{T}$ by some subgroup of the automorphism group Aut $\mathcal{T}$. To identify Aut $\mathcal{T}$ (and Aut $\mathcal{F}$) we need:

Lemma 4.3. *$\mathcal{T}$ is the only triangular embedding of $\mathcal{F}$.*

Proof. It is sufficient to show that each edge $e = vw$ of $\mathcal{F}$ is contained in exactly two triangular circuits of $\mathcal{F}$, since these must correspond to the faces in any triangular embedding. By Proposition 3.1 (iii) we can assume that $v = 0$ and $w = \infty$, in which case it follows easily from the condition $ry - sx = \pm 1$ that 1 and -1 are the only vertices adjacent to both v and w.

Corollary 4.4. Aut $\mathcal{F}$ = Aut $\mathcal{T}$ = PGL(2,$\mathbb{Z}$).

Proof. Clearly Aut $\mathcal{T} \leq$ Aut $\mathcal{F}$, since any automorphism of $\mathcal{T}$ must leave invariant its 1-skeleton $\mathcal{F}$. On the other hand, Lemma 4.3 implies that any automorphism of $\mathcal{F}$ must preserve $\mathcal{T}$ (since it permutes the triangles in $\mathcal{F}$), so we have Aut $\mathcal{F} \leq$ Aut $\mathcal{T}$, and hence Aut $\mathcal{F}$ = Aut $\mathcal{T}$.

Now the *extended modular group* Π = PGL(2,Z) consists of Γ together with the transformations

$$z \mapsto \frac{a\bar{z} + b}{c\bar{z} + d} \qquad (a,b,c,d \in \mathbb{Z};\ ad - bc = -1),$$

which induce orientation-reversing isometries of $\mathcal{U}$; we have $\Pi = \Gamma \cup r\Gamma$, where r is the reflection $z \mapsto -\bar{z}$. Since r and Γ induce automorphisms of $\mathcal{F}$, we have $\Pi \leq$ Aut $\mathcal{F}$.

Conversely, let $g \in \mathrm{Aut}\,\mathcal{F}$; then $g(01\infty)$ is a triangle of $\mathcal{F}$, and one easily verifies that some $h \in \Pi$ takes $g(0)$, $g(1)$ and $g(\infty)$ to 0,1 and ∞ respectively. Thus hg is an automorphism of $\mathcal{T}$ fixing 0,1 and ∞, so it fixes the triangle 01∞; it therefore fixes the triangles (such as $\infty 12$) adjacent to 01∞, and hence the triangles adjacent to these, and so on. It follows that hg is the identity automorphism, so

$$g = h^{-1} \in \Pi$$

and hence $\mathrm{Aut}\,\mathcal{F} \leq \Pi$. Thus $\mathrm{Aut}\,\mathcal{F} = \Pi$, as required.

By the universal nature of $\mathcal{T}$ (see [11]) we have the remarkable consequence that triangular maps are classified by the conjugacy classes of subgroups of PGL(2,Z) (see also the comment preceding Lemma 5.3 for a generalisation of this).

Finally, we relate $\mathcal{F}$ to continued fractions. It is easily seen that $\mathcal{F}$ is connected: for instance any pair of rational vertices are joined by a path in $\mathcal{F}_m$ (for sufficiently large m), while ∞ is adjacent to 0. We can therefore define the *Farey distance* $d(v,w)$ between any two vertices $v, w \in \hat{\mathbf{Q}}$ to be the minimum number of edges in any path from v to w in $\mathcal{F}$; thus d is a metric on $\hat{\mathbf{Q}}$, and Γ is a group of isometries. The vertices w with $d(\infty,w) = 1$ are the integers, while at distance 2 from ∞ we have the rationals

$$p + q^{-1} \quad (p,q \in \mathbf{Z},\ |q| \geq 2).$$

In [13], the third author showed how a shortest path in $\mathcal{F}$ from ∞ to w can be found by expressing w as a continued fraction

$$w = c_1 - \cfrac{1}{c_2 - \cfrac{1}{c_3 - \ddots - \cfrac{1}{c_m}}},$$

$(c_i \in \mathbf{Z})$, the distance $d(\infty,w)$ being equal to m. For example,

$$\frac{6}{11} = 1 - \cfrac{1}{2 - \cfrac{1}{-5}},$$

corresponding to the path of length 3 given by

$$\infty \to 1 \to 1 - \frac{1}{2} = \frac{1}{2} \to 1 - (2 - (-5)^{-1})^{-1} = 6/11.$$

If $w = x/y$ then $d(\infty,w)$ is also the number of divisions involved in using the least remainders algorithm to find the highest common factor (x,y).

For example:

$6 = 1.\ 11 - 5$ (with $|5| \leq \frac{1}{2}|11|$),

$11 = -2.\ -5 + 1$ (with $|1| \leq \frac{1}{2}|-5|$),

$-5 = -5.1$ (with remainder 0).

Thus we can regard $d(\infty,w)$ as a measure of the "complexity" of w.

5. The graphs $\mathcal{G}_{u,n}$ and $\mathcal{F}_{u,n}$

In this final section we see how the properties of $\mathcal{F} = \mathcal{G}_{1,1}$ extend to the other suborbital graphs $\mathcal{G}_{u,n}$.

As we saw when proving Corollary 3.5, each $\mathcal{G}_{u,n}$ is a disjoint union of $\psi(n)$ subgraphs, the vertices of each subgraph forming a single block with respect to the Γ-invariant equivalence relation $\underset{n}{\approx}$ defined by $ry - sx \equiv 0 \bmod n$. Since Γ acts transitively on $\hat{\mathbb{Q}}$, it permutes these blocks transitively, so the subgraphs are all isomorphic. We let $\mathcal{F}_{u,n}$ be the subgraph of $\mathcal{G}_{u,n}$ whose vertices form the block

$$[\infty] = \{x/y \in \hat{\mathbb{Q}} \mid y \equiv 0 \bmod n\}$$

containing ∞, so that $\mathcal{G}_{u,n}$ consists of $\psi(n)$ disjoint copies of $\mathcal{F}_{u,n}$. Theorem 3.2 immediately gives:

Theorem 5.1. $r/s \rightarrow x/y$ *in* $\mathcal{F}_{u,n}$ *if and only if either*

(a) $x \equiv ur \bmod n$ *and* $ry - sx = n$, *or*

(b) $x \equiv -ur \bmod n$ *and* $ry - sx = -n$.

By our general discussion of imprimitivity in Section 2, the subgroup of Γ leaving $\mathcal{F}_{u,n}$ invariant is the congruence subgroup $\Gamma_0(n)$ inducing $\underset{n}{\approx}$ on $\hat{\mathbb{Q}}$. Thus $\Gamma_0(n) \leq \mathrm{Aut}\ \mathcal{F}_{u,n}$.

Theorem 5.2. $\Gamma_0(n)$ *permutes the vertices and the edges of* $\mathcal{F}_{u,n}$ *transitively.*

Proof. Let v, w be vertices of $\mathcal{F}_{u,n}$. By Proposition 3.1(ii), $w = g(v)$ for some $g \in \Gamma$. Now Γ permutes the blocks, and v,w both lie in the block $[\infty]$, so g preserves $[\infty]$ and hence lies in $\Gamma_0(n)$. The proof for edges is similar.

In the simplest case (n=1) we have

$$\mathfrak{F}_{1,1} = \mathfrak{G}_{1,1} = \mathfrak{F}.$$

The next simplest graph, $\mathfrak{G}_{1,2}$ is shown in Fig. 3: it consists of $\psi(2) = 3$ isomorphic copies of $\mathfrak{F}_{1,2}$, with unbroken, broken and dotted edges (represented as hyperbolic geodesics in $\mathcal{U}$) indicating the blocks $[\infty]$, $[0]$, $[1]$ which consist of the elements $r/s \in \hat{\mathbb{Q}}$ with s even, r even, or both odd.

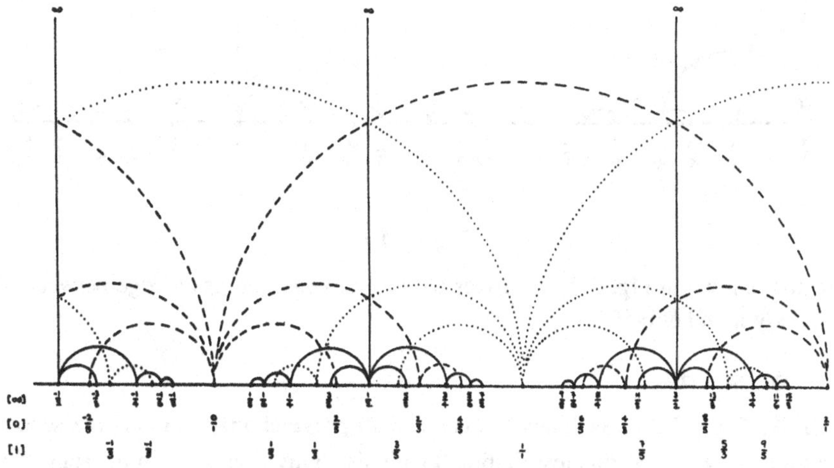

Fig. 3: $\mathfrak{G}_{1,2}$

By Corollary 3.4, $\mathfrak{G}_{1,2}$ is undirected, and hence so is $\mathfrak{F}_{1,2}$. This

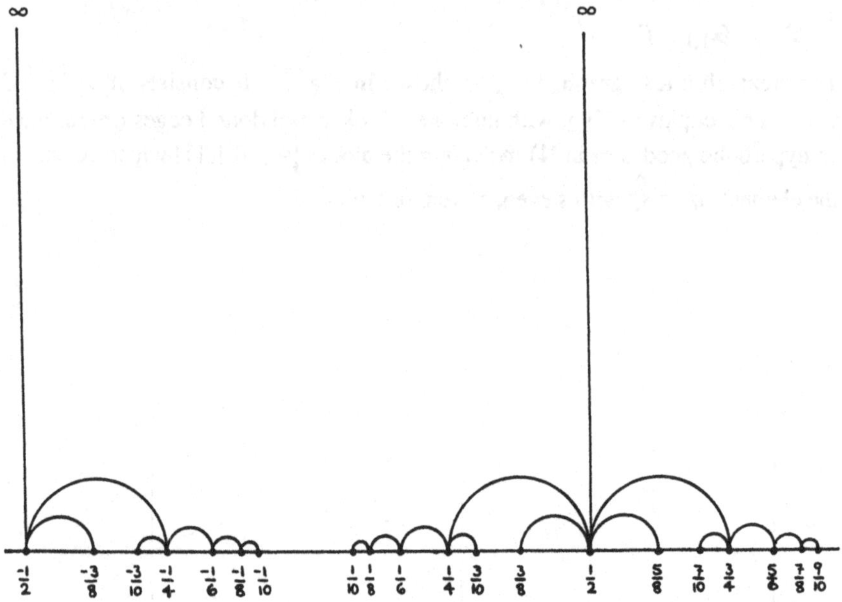

Fig. 4: $\mathfrak{F}_{1,2}$

graph, shown in Fig. 4, has vertices r/s with s even; there is an edge between r/s and x/y if and only if

$$ry - sx = \pm 2.$$

Unlike $\mathfrak{G}_{1,2}$, $\mathfrak{F}_{1,2}$ tessellates $\mathcal{U}$ (as we shall prove shortly); in [11] the second author showed that this tessellation $\mathcal{T}_{1,2}$ is the "universal map" in the sense that any map on a surface is isomorphic to a quotient of $\mathcal{T}_{1,2}$ by some group of automorphisms. Thus maps are classified by conjugacy classes of subgroups of Aut $\mathcal{T}_{1,2}$, which is the extension of $\Gamma_0(2)$ by the reflection $z \mapsto -\bar{z}$.

The following useful lemma follows easily from Theorem 5.1:

Lemma 5.3. (i) *There is an isomorphism* $\mathfrak{F}_{u,n} \to \mathfrak{F}_{-u,n}$, *given by* $v \mapsto -v$ *for all vertices* v *of* $\mathfrak{F}_{u,n}$.

(ii) *If* $m|n$ *then there is an isomorphism from* $\mathfrak{F}_{u,n}$ *to a subgraph of* $\mathfrak{F}_{u,m}$, *given by* $v \mapsto nv/m$ *for all vertices v of* $\mathfrak{F}_{u,n}$.

Taking m = 1 in (ii), we obtain:

Corollary 5.4. *There is an isomorphism from* $\mathfrak{F}_{u,n}$ *to a subgraph of* $\mathfrak{F}$, *given by* $f : v \mapsto nv$ *for all vertices* v *of* $\mathfrak{F}_{u,n}$.

As in Figs. 2 and 4, we can obtain a representation $\mathcal{T}_{u,n}$ of $\mathfrak{F}_{u,n}$ in $\mathcal{U}$ by using hyperbolic geodesics as edges. Since $\Gamma_0(n)$ preserves $\mathfrak{F}_{u,n}$ and acts as a group of isometries of $\mathcal{U}$, it leaves $\mathcal{T}_{u,n}$ invariant. The function

$$f : \mathcal{U} \to \mathcal{U},$$
$$z \mapsto nz,$$

in Corollary 5.4 sends geodesics to geodesics, so it sends $\mathcal{T}_{u,n}$ to a subset of the Farey tessellation $\mathcal{T} = \mathcal{T}_{1,1}$. By Corollary 4.2, it follows that no edges of $\mathfrak{F}_{u,n}$ cross in $\mathcal{U}$, so we have:

Theorem 5.5. $\mathfrak{F}_{u,n}$ *is embedded in* $\mathcal{U}$, *that is,* $\mathcal{T}_{u,n}$ *is a tessellation of* $\mathcal{U}$.

(By transitivity of Γ, the same is true for each of the remaining $\psi(n) - 1$ copies of $\mathfrak{F}_{u,n}$ contained in $\mathfrak{G}_{u,n}$.)

Theorem 5.6. $\mathfrak{F}_{1,2}$ *is connected.*

Proof. It is sufficient to show that each vertex v of $\mathfrak{F}_{1,2}$ is joined to ∞ by a path in $\mathfrak{F}_{1,2}$. Putting $v = a/2b$, where $a, b \in \mathbb{Z}$, $b \geq 1$, and $(a,2b) = 1$, we will argue by induction on b. If b=1 then v and ∞ are adjacent, so assume that $b \geq 2$, and that the result is valid for all denominators less than 2b. We will show that v is adjacent to a vertex w with smaller denominator; by the induction hypothesis, w is joined by a path to ∞, and hence so is v.

Since (a,b) = 1 there exist integers c, d with ad - bc = 1. Replacing c and d by c+ka and d+kb, for some suitable $k \in \mathbb{Z}$, we can assume that $0 < d < b$. If c is odd, then $w = c/2d$ is the required vertex of $\mathfrak{F}_{1,2}$, adjacent to v since

$$a.2d - 2b.c = 2.$$

If c is even then a-c is odd, and since $0 < b - d < b$ we can take

$$w = (a-c)/2(b-d),$$

adjacent to v because $a.2(b-d) - 2b.(a-c) = 2(bc-ad) = -2$.

Corollary 5.7. $\mathfrak{G}_{1,2}$ *has* $\psi(2) = 3$ *connected components.*

Similar (and only slightly more complicated) arguments show that $\mathfrak{F}_{1,n}$ is connected for all $n \leq 4$; in each case, $\mathbb{Z}_n$ is sufficiently small that, given any vertex v, we can find a suitable vertex w (if n > 4, there are too many possibilities for c and d mod n). In particular, we have an alternative proof that $\mathfrak{F}$ is connected. By Lemma 5.3(i), $\mathfrak{F}_{2,3}$ and $\mathfrak{F}_{3,4}$ are also connected, so we have:

Theorem 5.8. $\mathfrak{F}_{u,n}$ *is connected for all* $n \leq 4$.

By contrast, we now prove:

Theorem 5.9. *If* $n = 5$, *then* $\mathfrak{F}_{u,n}$ *is not connected.*

Proof. First consider $\mathfrak{F}_{1,5}$, a directed graph by Corollary 3.4. Here $r/s \to x/y$ if and only $x \equiv r \bmod 5$ and $ry - sx = 5$ or $x \equiv -r \bmod 5$ and $ry - sx = -5$. Thus vertices r/s with $r \equiv \pm 1 \bmod 5$ are never joined to vertices x/y with $x \equiv \pm 2 \bmod 5$, so they lie in distinct connected components, and $\mathfrak{F}_{1,5}$ is not connected. By Lemma 5.3(i), the same holds for $\mathfrak{F}_{4,5}$.

Now consider $\mathfrak{F}_{2,5}$ (see Fig. 5); by Corollary 3.4,

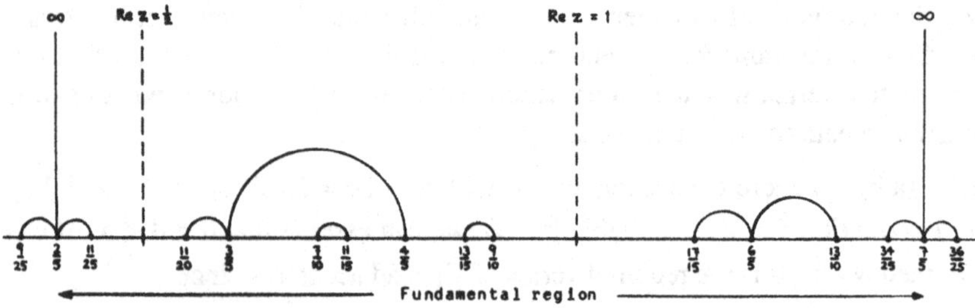

Fig. 5: $\mathfrak{F}_{2,5}$

this is an undirected graph. The embedding $\mathcal{T}_{2,5}$ of $\mathfrak{F}_{2,5}$ in $\mathcal{U}$ is invariant under the translation $z \mapsto z+1$, so we can regard the strip $2/5 \leq \text{Re } z \leq 7/5$ as a fundamental region for $\mathcal{T}_{2,5}$ in the sense that $\mathcal{T}_{2,5}$ consists of infinitely many translates of this strip. We note that in $\mathfrak{F}_{2,5}$, ∞ is adjacent to 2/5 and 7/5, but to no intermediate vertices.

We shall show that no vertices of $\mathcal{F}_{2,5}$ between $\frac{1}{2}$ and 1 are adjacent to vertices outside this interval, so the graph has at least two components (in fact, infinitely many, since the same applies to each interval $(k-\frac{1}{2}, k)$, $k \in \mathbf{Z}$). Suppose that an edge crosses $\mathrm{Re}\, z = \frac{1}{2}$, so it joins $v = a/5b$ to $w = c/5d$, where $v < \frac{1}{2} < w$. By Corollary 5.4 (with $n = 5$), $5v$ and $5w$ must be adjacent in $\mathcal{F}$, and hence adjacent in some $\mathcal{F}_m$ by Lemma 4.1. Now

$$5v < 5/2 < 5w,$$

with $5/2 \in \mathcal{F}_2$, so we must have $m = 1$ with $5v = 2$ and $5w = 3$, that is, $v = 2/5$ and $w = 3/5$. However, $3 \not\equiv \pm 2.2 \bmod 5$, so 2/5 and 3/5 are *not* adjacent in $\mathcal{F}_{2,5}$.

A similar argument shows that no edge crosses $\mathrm{Re}\, z = 1$, and since vertices between $\frac{1}{2}$ and 1 are not adjacent to ∞, it follows that $\mathcal{F}_{2,5}$ is not connected. By Lemma 5.3(i), the same conclusion holds for $\mathcal{F}_{3,5}$.

Similar but rather tedious arguments (in some cases using the lines $\mathrm{Re}\, z = \frac{1}{3}$ and $\mathrm{Re}\, z = \frac{2}{3}$ in place of $\mathrm{Re}\, z = \frac{1}{2}$) can be used to extend Theorems 5.8 and 5.9 to the following general result:

Theorem 5.10. *$\mathcal{F}_{u,n}$ is connected if and only if $n \leq 4$.*

As remarked in the proof of Theorem 5.9, $\mathcal{F}_{2,5}$ has infinitely many connected components (in fact, $\aleph_0$ components, since it has countably many vertices); since $\Gamma_0(n)$ permutes the vertices transitively, these components are all isomorphic. It follows that Aut $\mathcal{F}_{2,5}$ is an uncountable group, since any permutation of the components is induced by some graph-automorphism. On the other hand Aut $\mathcal{T}_{2,5}$ is countable, since $\mathcal{T}_{2,5}$ has countably many edges and at most 4 map automorphisms can leave an edge invariant. This is in contrast with the situation in Section 4, where Aut $\mathcal{F}$ = Aut $\mathcal{T}$.

Finally, we consider circuits in the graphs $\mathcal{F}_{u,n}$. The Farey graph $\mathcal{F} = \mathcal{F}_{1,1}$ contains triangles (see Fig. 2), while the directed graph $\mathcal{F}_{1,3}$ contains directed triangles such as

$$\infty \rightarrow \frac{1}{3} \rightarrow \frac{2}{3} \rightarrow \infty.$$

(By a *directed triangle* we mean three vertices v_1, v_2, v_3 such that

$$v_1 \rightarrow v_2 \rightarrow v_3 \rightarrow v_1;$$

an *undirected triangle* will denote the configuration $v_1 \to v_2 \leftarrow v_3 \to v_1$; in a self-paired suborbital graph, the two concepts are equivalent.)

Theorem 5.11. (i) $\mathfrak{F}_{u,n}$ *contains directed triangles if and only if* $u^2 \pm u + 1 \equiv 0$ mod n.

(ii) *If* $n > 1$, *then* $\mathfrak{F}_{u,n}$ *contains no anti-directed triangles.*

Proof. (i) Suppose that $\mathfrak{F}_{u,n}$ contains a directed triangle. By Theorem 5.2 we may suppose that this has the form $\infty \to u/n \to v \to \infty$ for some vertex v of $\mathfrak{F}_{u,n}$. By applying Theorem 5.1 to the edge $v \to \infty$, we see that $v = r/n$ for some integer r, and by applying it to $u/n \to v = r/n$ we see that either $r \equiv u^2$ mod n and $u - r = 1$, giving $u^2 - u + 1 \equiv 0$ mod n, or $r \equiv -u^2$ mod n and $u - r = -1$, giving

$$u^2 + u + 1 \equiv 0 \text{ mod n.}$$

Conversely, if $u^2 \pm u + 1 \equiv 0$ mod n then Theorem 5.1 implies that there is a directed triangle

$$\infty \to u/n \to (u\pm1)/n \to \infty \text{ in } \mathfrak{F}_{u,n}.$$

(ii) As in (i), we may assume that an anti-directed triangle has the form

$$\infty \to u/n \leftarrow r/n \to \infty$$

for some integer r. By applying Theorem 5.1 to the middle edge, we see that either $r \equiv 1$ mod n and $r - u = 1$, or $r \equiv -1$ mod n and $r - u = -1$. In either case, $u \equiv 0$ mod n, which is impossible for $n > 1$ since u is a unit mod n.

For example, $\mathfrak{F}_{1,3}$ contains the triangle

$$\infty \to 1/3 \to 2/3 \to \infty.$$

(This is a face of $\mathfrak{T}_{1,3}$; however, the three adjacent faces all have infinitely many edges!)

Since a self-paired suborbital graph $\mathfrak{G}_{u,n}$ satisfies $u^2 + 1 \equiv 0$ mod n, Theorem 5.11 implies that it cannot contain triangles if $n > 1$. In fact, we can prove rather more:

Theorem 5.12. *If* $n > 1$, *then every self-paired suborbital graph* $\mathfrak{G}_{u,n}$ *is a forest.*

Proof. Since $\mathfrak{G}_{u,n}$ is a disjoint union of isomorphic copies of $\mathfrak{F}_{u,n}$, it is sufficient to show that $\mathfrak{F}_{u,n}$ contains no circuits. If there is a circuit, then by Theorem 5.2 we may assume it has the form

$\infty \to v_1 \to \ldots \to v_k \to \infty$

(with all k+1 vertices distinct). Since $\mathcal{F}_{u,n}$ is self-paired, Theorem 5.1 and Corollary 3.4 give $v_1 = a_1/n$ and $v_k = a_k/n$ where $a_1, a_k \equiv u \bmod n$, so $v_1 - v_k \in \mathbb{Z}$. Since $n > 1$, we have $v_1, v_k \notin \mathbb{Z}$, so some integer m lies between them. It follows that some edge $v_i \to v_{i+1}$ of $\mathcal{F}_{u,n}$ crosses the line Re z = m in $\mathcal{U}$, so by Corollary 5.4 the edge $nv_i \to nv_{i+1}$ of $\mathcal{F}$ crosses Re z = nm. However, this contradicts Corollary 4.2, since Re z = nm is an edge of $\mathcal{F}$.

Corollary 5.13. $\mathcal{G}_{1,2}$ *is a forest.*

Corollary 5.14. *If* n *is even then* $\mathcal{G}_{u,n}$ *is a forest.*

Proof. This follows immediately from Corollary 5.13 and Lemma 5.3(ii) (with m = 2).

We conjecture that $\mathcal{G}_{u,n}$ is a forest if and *only if* it contains no triangles, that is, if and only if $u^2 \pm u + 1 \not\equiv 0 \bmod n$.

References

1. N L Biggs, *Graphs with large girth*, preprint.
2. N L Biggs & A T White, *Permutation groups and combinatorial structures* (London Math. Soc. Lecture Notes **33**, Cambridge University Press, Cambridge, 1979).
3. G H Hardy & E M Wright, *An introduction to the theory of numbers*, 5th ed. (Oxford University Press, Oxford, 1979).
4. A Hurwitz, Über die Reduktion der binären quadratischen Formen, *Math. Ann.* **45** (1894), 85-117.
5. G A Jones & D Singerman, *Complex functions: an algebraic and geometric viewpoint* (Cambridge University Press, Cambridge, 1987).
6. W J LeVeque, *Fundamentals of number theory* (Addison-Wesley, Reading, Mass., 1977).
7. P M Neumann, Finite permutation groups, edge-coloured graphs and matrices, in *Topics in group theory and computation* (Ed. M P J Curran, Academic Press, London, New York, San Francisco, 1977).
8. B Schoeneberg, *Elliptic modular functions* (Springer-Verlag, Berlin, Heidelberg, New York, 1974).
9. C Series, The geometry of Markoff numbers, *Math. Intelligencer* **7** (1985), 20-29.
10. C C Sims, Graphs and finite permutation groups, *Math. Z.* **95** (1967), 76-86.
11. D Singerman, Universal tessellations, *Revista Matemática de la Universidad Complutense de Madrid* **1** (1988), 111-123.

12. T Tsuzuku, *Finite groups and finite geometries* (Cambridge University Press, Cambridge, 1982).

13. K Wicks, *The Farey graph and related topics* (Project, Faculty of Mathematical Studies, University of Southampton, 1983).

ON THE N-CENTRE OF A GROUP

LUISE-CHARLOTTE KAPPE

SUNY at Binghamton, New York 13901, USA

MARTIN L NEWELL

University College at Galway, Ireland

1. Introduction

Let n be an integer. Two elements x, y in a group G n-commute if

$$(xy)^n = x^n y^n \text{ and } (yx)^n = y^n x^n ,$$

see [2]. A group is n-abelian if any two elements n-commute. In [1], R Baer introduced the n-centre Z(G,n) of a group G as the set of those elements which n-commute with every element in the group. However, $(ax)^n = a^n x^n$ for all $x \in G$ implies $(xa)^n = x^n a^n$ for all $x \in G$, and vice versa, as we will show in the next section. Thus, the set of elements which n-commute with all elements in the group from one side only is already all of Z(G,n).

The n-centre which can readily be seen to be a characteristic subgroup, shares many properties with the centre, some of which already have been explored in [1] and [2]. For instance, if the central quotient of a group is (locally) cyclic, then the group is abelian. Similarly, it follows by Corollary 1 in [2] that a group is n-abelian, if the quotient modulo its n-centre is (locally) cyclic. The topic of this paper is to shed further light on these similarities by investigating various characterizations and embedding properties of the n-centre.

The centre can be characterized as the margin of the commutator word [x,y]. As we will see in this paper, the n-centre can be characterized likewise as the margin of the n-commutator word $(xy)^n y^{-n} x^{-n}$. This result yields some interesting connections with a conjecture of P Hall on margins.

It can be easily seen that the 2-centre of a group coincides with its centre. In the fourth and fifth sections of this paper we will discuss characterizations and embeddings in the upper central series for other values of n. Since any group of exponent n is equal to its n-centre, such results can only be expected for special values of n or restrictions on the class of groups under consideration.

Obviously, the 2-centre is contained in the n-centre for any integer n. Already Baer observed in [1] that always $Z(G,n) = Z(G,1-n)$. Based on results in [9], we will give a complete characterization of those integers m as a function of n such that $Z(G,n) \subseteq Z(G,m)$ for any group G.

It is well known that a group which is n-abelian for 3 consecutive integers is abelian. Similarly, it can be shown that the intersections of the n-centres for 3 consecutive integers is equal to the centre. More generally, we will give a characterization of a set of integers S such that the intersection of the n-centres for all n in S is equal to the centre.

2. Basic results

Let

$$S_1(G,n) = \{a \in G \mid (ax)^n = a^n x^n \ \forall\, x \in G\}$$

and

$$S_2(G,n) = \{a \in G \mid (xa)^n = x^n a^n \ \forall\, x \in G\}.$$

According to [1], the n-centre is defined as

$$Z(G,n) = S_1(G,n) \cap S_2(G,n).$$

However, we will show $S_1(G,n) = S_2(G,n)$. Thus only one of the n-commutativity conditions suffices to define the n-centre.

Theorem 2.1. *For an integer* n *and any group* G

$$Z(G,n) = \{a \in G \mid (ax)^n = a^n x^n \ \forall\, x \in G\} = \{a \in G \mid (xa)^n = x^n a^n \ \forall\, x \in G\}.$$

Proof. For $a \in S_1(G,n)$ it follows by inversion that $(x^{-1}a^{-1})^n = x^{-n}a^{-n}$ for all $x \in G$, and hence $(ya^{-1})^n = y^n a^{-n}$ for all $y \in G$. Thus

$$ax^n a^{-1} = (axa^{-1})^n = a^n (xa^{-1})^n = a^n x^n a^{-n},$$

and we conclude $[a^{n-1}, x^n] = 1$. But

$$(ax)^n = a(xa)^{n-1}x = a^n x^n,$$

so

$$(xa)^{n-1} = a^{n-1}x^{n-1}.$$

Therefore $(xa)^n = (xa)^{n-1}(xa) = a^{n-1}x^n a = x^n a^n$, and hence $S_1(G,n) \subseteq S_2(G,n)$. A similar argument shows $S_2(G,n) \subseteq S_1(G,n)$.

The following lemma collects various facts about the elements in the n-centre, some of which already can be found in [1]. Observe throughout that $Z(G,1) = Z(G,0) = G$.

Lemma 2.2. *Let* $a \in Z(G,n)$. *Then*

(i) $[a^{n-1},x^n] = 1$ *for all* $x \in G$;

(ii) $a \in Z(G,1-n)$;

(iii) $[a^n,x] = [a,x]^n = [a,x^n]$ *for all* $x \in G$;

(iv) $1 = [a,x^{n(1-n)}] = [a^{n(1-n)},x] = [a,x]^{n(1-n)} = [a^n,x^{1-n}]$ *for all* $x \in G$;

(v) $a^n \in Z(G,n-1)$.

Proof. (i) This was shown for $a \in S_1(G,n)$ in Theorem 2.1. Since $S_1(G,n) = Z(G,n)$ by the same theorem, the claim follows.

(ii) By Theorem 2.1 $(xa)^{n-1} = a^{n-1}x^{n-1}$ for $a \in S_1(G,n)$. Inversion and Theorem 2.1 lead to the desired result.

(iii) We observe $[a,x]^n = (a^{-1}a^x)^n = a^{-n}a^{nx} = [a^n,x]$ for the first equality. For the second part observe that a^nx^{-n} commutes with ax^{-1}, since $a^nx^{-n} = (ax^{-1})^n$. Thus $a^{n+1}[a,x^n]x^{-n-1} = a^nx^{-n}ax^{-1} = ax^{-1}a^nx^{-n} = a^{n+1}[a^n,x]x^{-n-1}$, and the claim follows.

(iv) By (i), (ii) and (iii) it follows

$$1 = [a^{1-n},x^n] = [a^{1-n},x]^n = [a,x]^{n(1-n)} = [a^{n(1-n)},x],$$

and similarly $1 = [a^{1-n},x^n] = [a,x^{n(1-n)}] = [a^n,x^{1-n}]$.

(v) We observe that by (iv) a^n and x^{n-1} commute. Thus

$$(a^nx)^{n-1} = (a^nx)^nx^{-1}a^{-n} = a^{n^2}x^{n-1}a^{-n} = a^{n(n-1)}x^{n-1}.$$

Motivated by the obvious characterization of the centre as those $a \in G$ for which $\langle a, x\rangle$ is abelian for all $x \in G$, we give here an analogue characterization of the n-centre.

Theorem 2.3. *Let* n *be an integer and* G *a group. Then* $a \in Z(G,n)$ *if and only if* $\langle a, x\rangle$ *is* n-*abelian for all* $x \in G$.

Proof. One direction is obvious. Corollary 1 in [2] states that $G/Z(G,n)$ cyclic implies $G = Z(G,n)$. Thus $\langle x,Z(G,n)\rangle$ is n-abelian for all $x \in G$. Hence $\langle a, x\rangle$ is n-abelian for all $a \in Z(G,n)$ and all $x \in G$.

The following lemma, needed in the proof of Theorem 5.1, is a simple consequence of the above theorem.

Lemma 2.4. *Given* $x \in G$ *and* $a \in Z(G,n)$. *Let* $X,Y \in \langle a^{n-1},x\rangle$ *with*

$$X = \prod_{\rho=1}^{r} a^{(n-1)i_\rho}x^{j_\rho} \text{ and } Y = \prod_{\rho=1}^{r} a^{i_\rho}x^{(n-1)j_\rho},$$

where $i_\rho, j_\rho \in \mathbf{Z}$. *Then*

$$X^n = a^{n(n-1)s}x^{nt} \text{ and } Y^n = a^{n(n-1)t}x^{ns}$$

with

$$s = \sum_{\rho=1}^{r} i_\rho \text{ and } t = \sum_{\rho=1}^{r} j_\rho.$$

Proof. We observe that $\langle a^{n-1},x\rangle$ is n-abelian by Theorem 2.3 and $a^{n(n-1)} \in Z(G)$ by Lemma 2.2 (iv), and our claim follows.

3. The n-commutator margin

Margins were introduced by P Hall in [4]. For a word $\psi(x_1,\ldots,x_k)$ in the variables $x_1,\ldots,x_k$ we define for any group G

$$\psi_i^*(G) = \{a \in G \mid \psi(g_1,\ldots,ag_i,\ldots,g_k) = \psi(g_1,\ldots,g_i,\ldots,g_k) \ \forall \ g_1,\ldots,g_k \in G\}$$

as the i-th partial margin of ψ, and

$$\psi^*(G) = \bigcap_{i=1}^{k} \psi_i^*(G)$$

as the margin of ψ in G. Partial margins as well as the margin are characteristic subgroups in G. We observe that $a \in \psi_i^*(G)$ if and only if

$$\psi(g_1,\ldots,g_ia,\ldots,g_k) = \psi(g_1,\ldots,g_i,\ldots,g_k)$$

for all $g_1,\ldots,g_k \in G$. The word subgroup of ψ in G is defined as

$$\psi(G) = \langle \psi(g_i,\ldots,g_k) \mid g_1,\ldots,g_k \in G\rangle.$$

For an integer n we denote with $\beta_n(x,y) = (xy)^n y^{-n} x^{-n}$ the n-commutator word, and with

$$\beta^*_{n,1}(G) = \{a \in G \mid \beta_n(ax,y) = \beta_n(x,y)\ \forall\ x,y \in G\}$$

and

$$\beta^*_{n,2}(G) = \{a \in G \mid \beta_n(x,ay) = \beta_n(x,y) \forall x,y \in G\}$$

the first and second partial n-commutator margin. Thus

$$\beta^*_n(G) = \beta^*_{n,1}(G) \cap \beta^*_{n,2}(G)$$

is the n-commutator margin, and

$$\beta_n(G) = [G,G;n] = \langle\, (xy)^n y^{-n} x^{-n} \mid x,y \in G \,\rangle$$

is the n-commutator subgroup, already introduced by Grün in [3].

With these definitions we have the following result.

Theorem 3.1. *For an integer* n *and any group* G *we have*

$$Z(G,n) = \beta^*_{n,1}(G) = \beta^*_{n,2}(G) = \beta^*_n(G).$$

Proof. Setting $x = 1$ for $a \in \beta^*_{n,1}(G)$ yields $(ay)^n = a^n y^n$ for all $y \in G$, hence $\beta^*_{n,1}(G) \subseteq Z(G,n)$. For $a \in \beta^*_{n,2}(G)$ similarly $a \in Z(G,n)$, and further

$$(xy)^n y^{-n} x^{-n} = (xay)^n (ay)^{-n} x^{-n} = (xay)^n y^{-n} a^{-n} x^{-n} = (xay)^n y^{-n} (xa)^{-n}.$$

This means $\beta_n(xa,y) = \beta_n(x,y)$ for all $x,y \in G$; hence we conclude

$$\beta^*_{n,2}(G) \subseteq \beta^*_{n,1}(G).$$

Finally, let $a \in Z(G,n)$. Then

$$(xya)^n (ya)^{-n} x^{-n} = (xy)^n a^n \cdot a^{-n} y^{-n} x^{-n}, \text{ or } \beta_n(x,ya) = \beta_n(x,y)$$

for all $x,y \in G$, hence $Z(G,n) \subseteq \beta^*_{n,2}(G)$. Observing

$$\beta^*_n(G) = \beta^*_{n,1}(G) \cap \beta^*_{n,2}(G),$$

the conclusion of the theorem follows.

The following result is already implicit in the work of Schur.

Theorem 3.2. ([13, Theorem 4.12]) *If* G *is a group whose centre has finite index* m, *then* G' *is finite and* $(G')^m = 1$.

Based on this theorem P Hall [5] made the following conjecture: Let G be a group and ψ a word. If $G/\psi^*(G)$ is finite of order m, then $\psi(G)$ is finite and the exponent of $\psi(G)$ is an m-number. This conjecture has been proved for outer commutator words [14]. The second part of the conjecture is not true in general, as was shown by Kleĭman in [12]. He gives a word v and a group G such that $|G/v^*(G)| = p^2$, where p is an odd prime, but $|v(G)| = 2$. It is still an open question whether the first part of Hall's conjecture is true in general, i.e. the finiteness of $G/\psi^*(G)$ implies the finiteness of $\psi(G)$ for any word ψ.

For the n-centre there is the following analogue to Theorem 3.2.

Theorem 3.3. ([1, Satz 6]) *If* G *is a group whose* n*-centre has finite index* m, *then* $\beta_n(G)$ *is finite and has exponent dividing* $m^2n(n-1)$.

In view of Theorem 3.1 and Baer's result we have now the following corollary.

Corollary 3.4. *If* $G/\beta_n^*(G)$ *is finite of order* m, *then* $\beta_n(G)$ *is finite and the exponent of* $\beta_n(G)$ *divides* $m^2n(n-1)$.

This shows that the finiteness part of Hall's conjecture is true for the n-commutator word. But it is an open question on whether the exponent of $\beta_n(G)$ is always an m-number. This depends on whether the estimate for the exponent is sharp in Theorem 3.3.

4. Characterization of the n-centre

In this section we investigate the 3-centre in general and the p-centre in the case of metabelian p-groups. In [10], similar investigations were done in the case of power margins. For an integer n the n-power margin is defined as

$$M_n(G) = \{a \in G \mid (ax)^n = x^n \ \forall\, x \in G\},$$

or alternatively $M_n(G) = \{a \in Z(G,n) \mid a^n = 1\}$. Results, relevant in this conext, are as follows.

Theorem 4.1. ([10, Theorem 1]) *Let* G *be a group. Then*

$$M_3(G) = \{a \in R_2(G) \mid a^3 = 1\}$$

and $M_3(G) \subseteq Z_3(G)$.

Theorem 4.2. ([10, Theorem 2]) *Let* G *be a metabelian group and* p *a prime. Then*

$$M_p(G) = \{a \in R_{p-1}(G) \mid a^p = 1\}$$

and $M_p(G) \subseteq Z_p(G)$.

Here $R_m(G) = \{a \in G \mid [a,_m x] = 1 \ \forall\ x \in G\}$ denotes the set of right m-Engel elements, where

$$[x,_m y] = [[x,_{m-1} y],y] \text{ and } [x,_1 y] = [x,y].$$

In case of the n-centre we have similar results.

Theorem 4.3. *Let* G *be a group. Then* $Z(G,3) = \{a \in R_2(G) \mid a^3 \in Z(G)\}$ *and* $Z(G,3) \subseteq Z_3(G)$.

Proof. For $a \in Z(G,3)$ we have

$$(xa^{-1})^3 = x^{1+a+a^2}a^{-3} = x^{-3}a^{-3}.$$

Therefore $x^{a+a^2} = x^2$.

Replacing x by x^{-1} in the above and inversion lead to $x^{a^2+a} = x^2$, hence

$$x^{a^2+a} = x^{a+a^2}.$$

After conjugation with a^{-1} we conclude that $[x^a,x] = 1$, and hence $a \in R_2(G)$. By Lemma 2.2 (v) we have $a^3 \in Z(G,2) = Z(G)$. Therefore $Z(G,3) \subseteq Y$, where $Y = \{a \in R_2(G) \mid a^3 \in Z(G)\}$.

Conversely, assume $a \in Y$. Then

$$[a,x,x] = 1 = [a,x,a]$$

and

$$[a^3,x] = 1 = [a,x]^3.$$

Thus,

$$(ax^{-1})^3 = a^3[a,x]^3x^{-3} = a^3x^{-3}$$

for all $x \in G$, and hence $Y \subseteq Z(G,3)$. Now $a^3 \in Z(G)$ implies $[a,x,y,z]^3 = 1$. By [11] it follows that $[a,x,y,z]^4 = 1$. Thus $[a,x,y,z] = 1$ for all $x,y,z \in G$, hence $Z(G,3) \subseteq Z_3(G)$.

Contrary to the case of power margins, there is no exponent restriction for the elements of the n-centre. Thus the analogue of Theorem 4.2 for the p-centre holds only in metabelian p-groups.

Theorem 4.4. *Let* p *be a prime and* G *a metabelian* p-*group. Then*

$$Z(G,p) = \{a \in R_{p-1}(G) \mid a^p \in Z(G)\}$$

and $Z(G,p) \subseteq Z_p(G)$.

Before proceeding with the proof of this theorem, we state four lemmas.

Lemma 4.5. ([10, Lemma 3]) *Let* G *be a metabelian group. If* $a \in R_n(G)$ *is of order prime to* n!, *then* $a \in Z_{n+1}(G)$.

Lemma 4.6. *Let* G *be a metabelian* p-*group. Suppose for all* a, x ∈ G *with* $a^p \in Z(G)$ *we have*

$$1 = \prod_{i=1}^{p-1} [a,_i\, x,_{p-1-i}\, a]. \qquad (1)$$

Then $[a,_s\, x,_{t-1}\, a] = 1$ *for all natural numbers* s, t *with* $s+t \geq p$.

Lemma 4.7. *Let* G *be a metabelian* p-*group with* $a \in R_{p-1}(G)$ *and* $a^p \in Z(G)$. *Then* $[a,_{p-1-i}\, x,_i\, a] = 1$ *for* i = 1,...,p-2 *and all* x ∈ G.

The proofs of Lemmas 4.6 and 4.7 follow along the line of Lemmas 4 and 5 in [10], where similar statements are made in case G is a metabelian group and $a^p = 1$. It can be easily verified that the same claims are true under the new assumptions.

Finally, we list the following expansion formulas in metabelian groups which can be found in [7].

Lemma 4.8. *Let* u,v ∈ G, *a metabelian group, and let* m *be a positive integer. Then*

(i) $$[u,v^m] = \prod_{i=1}^{m} [u,_i\, v]^{\binom{m}{i}};$$

(ii) $$(uv^{-1})^m = u^m \left(\prod_{0<i+j<m} [u,_i\, v,_j\, u]^{\binom{m}{i+j+1}} \right) v^{-m}.$$

Proof of Theorem 4.4. Application of Lemma 4.8 (ii) for $x,a \in G$ and $m = p$ yields

$$(ax^{-1})^p = a^p \left(\prod_{0<i+j<p} [a,_i x,_j a]^{\binom{p}{i+j+1}} \right) \cdot x^{-p}. \tag{2}$$

First assume $a \in Z(G,p)$ and let $|a| = p^\alpha$. Then there exist integers λ and μ such that $\lambda(p-1) + \mu p^{\alpha-1} = 1$. By Lemma 2.2 (iv) we obtain

$$[a^p,x] = [a^{\lambda p(p-1)+\mu p^\alpha}, x] = [a^{\lambda p(p-1)},x] = [a^\lambda,x]^{p(p-1)} = 1.$$

Hence $a^p \in Z(G)$ and $[a,x]^p = 1$ for all $x \in G$. Thus (2) reduces to

$$1 = \prod_{i=1}^{p-1} [a,_i x,_{p-1-i} a].$$

We can now apply Lemma 4.6 and conclude that each factor in the above product equals 1, in particular $[a,_{p-1} x] = 1$. Hence $Z(G,p) \subseteq \{a \in R_{p-1}(G) \mid a^p \in Z(G)\}$.

Conversely, assume $a \in R_{p-1}(G)$ and $a^p \in Z(G)$. Since G is a p-group, Lemma 4.5 yields $a \in Z_p(G)$. Therefore

$$1 = [a^p,x_1,\ldots,x_{p-1}] = [a,x_1,\ldots,x_{p-1}]^p$$

for all $x_1,\ldots,x_{p-1} \in G$. By an inductive argument and Lemma 4.8 (i) we obtain $[a,x_1,\ldots,x_k]^p = 1$ for all $x_1,\ldots,x_k \in G$, $k \geq 1$. Thus (2) reduces to

$$(ax^{-1})^p = a^p \left(\prod_{i=1}^{p-2} [a,_i x,_{p-1-i} a] \right) \cdot x^{-p}.$$

Now we can apply Lemma 4.7 and obtain that each factor in the above product of commutators equals 1. Hence $(ax^{-1})^p = a^p x^{-p}$ for all $x \in G$. Thus $a \in Z(G,p)$.

The fact that $Z(G,p) \subseteq R_{p-1}(G)$ together with G being a p-group implies $Z(G,p) \subseteq Z_p(G)$ by Lemma 4.5.

5. The 4-centre of a group

A group of exponent 4 is not necessarily nilpotent. Since groups of exponent 4 coincide with their 4-power margin and their 4-centre, we cannot expect a general result on the embedding of $M_4(G)$ and $Z(G,4)$ into the upper central series. However, what is known about the structure of $B(2,4)$, the 2 generator Burnside group of exponent 4, allows us to say something about the structure of $\langle a, x\rangle$, where $x \in G$ and $a \in Z(G,4)$ or $M_4(G)$, respectively.

The group B(2,4) is nilpotent of class 5 precisely (see e.g. [8]). It follows that in the free group $\langle x,y\rangle$ every commutator of total weight 6 in x and y can be expressed as a product of fourth powers of elements in $\langle x,y\rangle$. In particular, Havas [6] gave an explicit formula for the 5-Engel word $[x,_5 y]$ involving a product of 296 fourth powers. However, our investigations only rely on the existence of such an identity, but not on its explicit form. We have the following results on the 4-centre and the 4-power margin.

Theorem 5.1. *For any group* G

$$Z(G,4) \subseteq \{a \in G \mid \gamma_6(\langle a,x\rangle) = 1 \;\; \forall\, x \in G\},$$

$$M_4(G) \subseteq \{a \in G \mid a^4 = 1 \text{ and } \gamma_6(\langle a,x\rangle) = 1 \;\; \forall x \in G\}.$$

The following corollary is an immediate consequence of the above theorem.

Corollary 5.2. *The elements in* Z(G,4) *and* $M_4(G)$ *are right and left 5-Engel elements, i.e.* $[a,_5 x] = [x,_5 a] = 1$ *for all* $x \in G$ *and* $a \in Z(G,4)$ *or* $M_4(G)$, *respectively.*

Proof of Theorem 5.1. We observe that $M_4(G) = \{a \in Z(G,4) \mid a^4 = 1\}$. Thus, it suffices to show that $\gamma_6(\langle x,a\rangle) = 1$ for $a \in Z(G,4)$ and all $x \in G$. Denote by C the set of commutators of weight 6 in x and y in the free group $\langle x, y\rangle$. Every c(x,y) in C can be expressed as a product of fourth powers of elements in $\langle x, y\rangle$, i.e.

$$c(x,y) = \prod_{\kappa=1}^{k} X_\kappa^4, \tag{3}$$

where $X_\kappa \in \langle x,y\rangle$, $\kappa = 1,\dots,k$. Substitution of a^3 for x, $a \in Z(G,4)$, into (3) and observing Lemma 2.4 as well as Lemma 2.2 (iv) yield

$$c(a^3,y) = \prod_{\kappa=1}^{k} a^{12s_\kappa} y^{4t_\kappa} = a^{12S}\, y^{4T} = c(a^3,1)\cdot c(1,y) = 1,$$

where $S = \sum_{\kappa=1}^{k} s_\kappa$ and $T = \sum_{\kappa=1}^{k} t_\kappa$. Hence $c(a^3,y) = 1$ for $a \in Z(G,4)$ and all $y \in G$. Similarly, by setting $y = a^3$, $a \in Z(G,4)$, we obtain

$$c(x,a^3) = a^{12T}\cdot x^{4S} = c(1,a^3)\cdot c(x,1) = 1$$

for all $x \in G$. By Lemma 2.2 (ii) we have $b \in Z(G,-3)$ for $b \in Z(G,4)$. Thus $[b^{-3},u] = [b,u]^{-3}$ for all $u \in G$ by Lemma 2.2 (iii). Since $Z(G,4)$ is normal in G, it follows that

$$1 = c(b^3,y) = (c(b^{-1},y))^{(-3)^\lambda},$$

where λ is the weight of x in c(x,t). Similarly,

$$1 = c(x,b^3) = c(x,b^{-1})^{(-3)^{6-\lambda}}.$$

Setting $b^{-1} = a$, it follows that the order of each c(a,x) or c(x,a) is a power of three.

Lemma 2.2 (v) implies $a^4 \in Z(G,3)$, hence $a^4 \in Z_3(G)$ by Theorem 4.3. Therefore

$$c(a^4,x) = c(x,a^4) = 1$$

for each $c \in C$. By Lemma 2.2 (iii) we have for $b \in Z(G,4)$ and any $u \in G$ that $[b^4,u] = [b,u]^4$. Thus the order of each c(a,x) or c(x,a) is a 2-power. This is a contradiction to our observations above, unless $c(a,x) = c(x,a) = 1$. We conclude therefore that $\gamma_6(\langle x,a\rangle) = 1$ for $a \in Z(G,4)$ and all $x \in G$.

6. The embedding of the n-centre into the m-centre

Obviously $Z(G,2) \subseteq Z(G,m)$ for all integers m and, as already stated in [1], $Z(G,n) = Z(G,1-n)$. In this section we will determine the set of those integers m such that for a given integer n and any group G we have $Z(G,n) \subseteq Z(G,m)$. We will characterize this set as the exponent semigroup of the free n-abelian group. To that end, we list here various definitions, notations and results from [9].

For any group G, the set of integers

$$\mathcal{E}(G) = \{n \in \mathbb{Z} \mid (xy)^n = x^n y^n \text{ for all } x,y \in G\}$$

is called the *exponent semigroup* of G, and can be characterized as the set of idempotent residues for certain modules $q_1,\dots,q_t$, called a Levi system and denoted by $B(q_1,\dots,q_t)$, where the q_i are integers with $|q_i| > 1$ and $(q_i,q_j) = 1$ for $i \neq j$. Specifically we have

$$B(q_1) = \{m = kq_1 + \delta \mid \delta = 0,1 \text{ and } k \in \mathbb{Z}\},$$

and

$$B(q_1,q_2) = \{m = kq_1 \cdot q_2 + \delta \mid \delta = 0,1,q_1,q_2 \text{ and } k \in \mathbb{Z}\}.$$

Let n be an integer and F a noncyclic free group with [F,F;n] its n-commutator subgroup. The factor group F/[F,F;n] is called a free n-abelian group. For its exponent semigroup we have the following result.

Theorem 6.1 ([9, Theorem 3]). *Let* F *be a free group with 2 or more generators. Then*

$$\mathfrak{E}(F/[F,F;n]) = \begin{cases} B(n,1-n) & \text{for } n \geq 4 \text{ or } n \leq -3 \\ B(3) & \text{for } n = 3 \text{ or } -2 \\ \mathbf{Z} & \text{for } n = 2 \text{ or } -1 \end{cases}$$

For the embedding of the n-centre into the m-centre we have the following result.

Theorem 6.2. *Let* n *be an integer* ≠ 0,1 *and* G *a group. Then* Z(G,n) ⊆ Z(G,m) *if and only if*

(i) m ∈ B(n,1-n) *for* n ≥ 4 *or* n ≤ -3,

(ii) m ∈ B(3) *for* n = 3 *or* -2,

(iii) m ∈ **Z** *for* n = 2 *or* -1.

Proof. (i) Let a ∈ Z(G,n) and n ≥ 4 or n ≤ -3. By Theorem 2.3 we obtain that ⟨a,x⟩ is n-abelian for all x ∈ G. Hence Theorem 6.1 implies that ⟨a,x⟩ is m-abelian for all m ∈ B(n,1-n). Another application of Theorem 2.3 yields now Z(G,n) ⊆ Z(G,m) for all m ∈ B(n,1-n).

Conversely, assume Z(G,n) ⊆ Z(G,m) for any group G. Consider G = F/[F,F;n]. Since G is n-abelian, we have G = Z(G,n). But Z(G,n) ⊆ Z(G,m) implies G = Z(G,m), and hence G is m-abelian. By Theorem 6.1 it follows that m ∈ B(n,1-n).

(ii) For a ∈ Z(G,3) = Z(G,-2) we have ⟨a,x⟩ is 3-abelian by Theorem 2.3. Thus ⟨a,x⟩ is m-abelian for all m ∈ B(3) by Theorem 6.1. We conclude Z(G,3) ⊆ Z(G,m) for m ∈ B(3). Conversely, assume Z(G,3) ⊆ Z(G,m) for any group G. Consider G = F/[F,F;3], a free 3-abelian group, thus G = Z(G,3). Again Z(G,3) ⊆ Z(G,m) implies G = Z(G,m) and G is m-abelian. By Theorem 6.1 it follows that m ∈ B(3).

(iii) We observe Z(G,2) = Z(G,-1) = Z(G) ⊆ Z(G,m) for any group G and all integers m.

It is well known that a group is abelian iff it is n-abelian for 3 consecutive integers. This follows from a more general result, which is a consequence of Theorem 2 in [9].

Theorem 6.3. *The following 2 conditions are equivalent for a group* G.

(i) G *is* n-*abelian for all* $n \in S$, *where* $S \subseteq \mathbf{Z}$ *with* $\gcd(n^2 - n; n \in S) = 2$.

(ii) G *is abelian.*

Similarly, it can be shown that the intersection of the n-centres for 3 consecutive integers is equal to the centre. This is an immediate corollary of the following theorem.

Theorem 6.4. *Let* S *be a subset of the integers with* $\gcd(n^2 - n; n \in S) = 2$. *Then*

$$\bigcap_{n \in S} Z(G,n) = Z(G).$$

Proof. For $a \in \bigcap_{n \in S} Z(G,n)$ we have by Theorem 2.3 that $\langle a,x \rangle$ is n-abelian for all $n \in S$ and all $x \in G$. Thus, Theorem 6.3 implies that $\langle a,x \rangle$ is abelian for all $x \in G$, hence $a \in Z(G)$. Since obviously

$$Z(G) \subseteq \bigcap_{n \in S} Z(G,n),$$

the conclusion of the theorem follows.

References

1. R Baer, Endlichkeitskriterien für Kommutatorgruppen, *Math. Ann.* **124** (1952), 161-177.
2. R Baer, Factorization of n-soluble and n-nilpotent groups, *Proc. Amer. Math. Soc.* **45** (1953), 15-26.
3. O Grün, Beiträge zur Gruppentheorie IV. Über eine charakteristische Untergruppe, *Math. Nachr.* **3** (1949), 77-94.
4. P Hall, Verbal and Marginal Subgroups, *J. Reine Angew. Math.* **182** (1940), 156-157.
5. P Hall, *Nilpotent Groups* (Canad. Math. Congress Summer Sem. Univ. Alberta, 1957).
6. G Havas, Commutators in groups expressed as products of powers, *Comm. Algebra* **9** (1981), 115-129.
7. G T Hogan & W P Kappe, On the H_p-problem for finite p-groups, *Proc. Amer. Math. Soc.* **20** (1969), 450-454.
8. B Huppert, *Endliche Gruppen I* (Berlin, 1965).
9. L C Kappe, On n-Levi groups, *Arch. Math.* **47** (1986), 198-210.
10. L C Kappe, On Power Margins, *J. Algebra* **122** (1989), 337-344.

11. W P Kappe, Die A-Norm einer Gruppe, *Illinois J. Math.* **5** (1961), 187-197.

12. Yu G Kleĭman, Some questions of the theory of varieties of groups, *Izv. Akad. Nauk. SSSR Ser. Mat.* **47** (1983), 37-74.

13. D J S Robinson, *Finiteness conditions and generalized soluble groups I*, (Berlin, 1972).

14. R F Turner-Smith, Marginal subgroup properties for outer commutator words, *Proc. London Math. Soc.* (3) **14** (1964), 321-341.

EXISTENTIALLY CLOSED FINITARY LINEAR GROUPS

OTTO H KEGEL & DIETER SCHMIDT

Albert-Ludwigs-Universität, D-7800 Freiburg i. Br., West Germany

The automorphism g of the vector space V over the (skew) field K is *finitary* if the fixed-point space $C_V(g)$ in V has finite codimension $\mathrm{codim}_K(V : C_V(g))$ in V. The set FGL(V) of all finitary automorphisms of V is a normal subgroup of the full linear group GL(V) of V. An abstract group G is *finitary linear* if it admits an embedding

$$\phi : G \to FGL(V)$$

for some vector space V over some (skew) field K. The stable groups GL(K) of Bass [1] are examples of rather special finitary linear groups. Recently some very interesting papers have been published on the structure of finitary linear groups (over commutative fields): [3], [4], [8].

For a subgroup G of FGL(V) there is an integer-valued degree function d associating with every finitely generated subgroup F of G the dimension of some finite-dimensional F-invariant subspace of V on which the action of F is concentrated. Imposing some of the properties of such a degree function on the numerical function d defined on the set of finitely generated subgroups of an abstract group G, one obtains the notion of an abstract degree function. Groups G with an abstract degree function d admit such an embedding into some FGL(V) that the degree function of FGL(V) coincides with d on the image of G. This characterisation of finitary linear groups is a slight variant of a result of J I Hall [4].

One observes that, for a suitable order relation, the class $\mathbf{C}$ of all pairs (G,d), G a group and d an abstract degree function on G, forms an inductive class. So every pair $(G,d) \in \mathbf{C}$ is contained in a pair (E,d) which is existentially closed in $\mathbf{C}$. In this note we determine to a large extent the structure of such existentially closed pairs (E,d): The groups E essentially have a local system consisting of stable groups over existentially closed (skew) fields.

Here the results of Mez [7] and their generalisation to the case of skew fields are most useful.

1. Degree Functions

For the finitely generated subgroup F of FGL(V) there are two important subspaces of V: the *fixed-point space*

$$C_V(F) := \{v \in V;\ v^f = v \text{ for all } f \in F\}$$

and the *commutator space*

$$[F,V] := \langle [f,v] := v^f - v;\ f \in F,\ v \in V \rangle.$$

The subspace $C_V(F)$ has finite codimension in V, the subspace [F,V] is finite-dimensional. Every subspace U of V containing [F,V] is invariant under the action of F. The action of F on V is *concentrated* in the F-invariant subspace U of V if

$$V = U + C_V(F),$$

i.e. if U has a complement in V on which F acts trivially; if

$$U \cap C_V(F) = (O)$$

then this complement ($C_V(F)$) is unique. The *degree* d(F) of F on V is defined as $\dim_K(U)$ where U is minimal among the subspaces of V on which the action of F is concentrated. This dimension is unique:

$$\dim_K[F,V] + \operatorname{codim}_K(V : ([F,V] + C_V(F))),$$

while the subspace U need not be unique. Let F_1 be another finitely generated subgroup of FGL(V) such that $F \subseteq F_1$, then one has $d(F) \leq d(F_1)$; if the action of F_1 on V is concentrated in the subspace U_1 of V then the action of F on U_1 is concentrated in a d(F)-dimensional subspace of U_1.

Lemma 1. *Let* $\mathfrak{U}$ *be an ultrafilter on the index set* I. *Let* $\{V_i,K_i\}_{i\in I}$ *be a family of vector spaces* V_i *over the (skew) field* K_i. *Suppose that for every* $i \in I$ *a group* G_i *of finitary automorphisms on* V_i *is given such that for every finitely generated subgroup* X *of* G_i *the degree* $d_i(X)$ *is bounded by the natural number* b *(independent of* i*). Then the ultraproduct*

$$G := \prod_{i\in I} G_i/\mathcal{U}$$

is a group of finitary automorphisms of the vector space $V := \prod_{i\in I} V_i/\mathcal{U}$ *over the (skew) field* $K := \prod_{i\in I} K_i/\mathcal{U}$, *and the degree function of the finitely generated subgroups of* G *is bounded by* b.

Proof. It is standard (e.g. [6, p.66]) that the group G acts faithfully on the vector space V. The vector space V_i has a decomposition $V_i = U_i + C_i$ where the action of G_i in V_i is concentrated in U_i and $\dim_{K_i}(U_i) \leq b$. But then the subspace

$$U := \prod_{i\in I} U_i/\mathcal{U}$$

of V is G-invariant of dimension at most b over K, the action of G on V is concentrated in U since the subspace $C := \prod_{i\in I} C_i/\mathcal{U}$ of V consists of fixed-points for G and supplements U.

The function d which associates with every finitely generated subgroup F of the abstract group G the natural number d(F) is an *abstract degree function on* G if for every finitely generated subgroup H of G there is a (skew) field K_H and a faithful linear representation ϕ_H of H into $GL(d(H),K_H)$ such that the restriction of the degree function of $GL(d(H), K_H)$ to the finitely generated subgroups of the image H^ϕ of H coincides with d.

The existence of an abstract degree function on a group characterises the finitary linear groups.

Theorem 1. *If the group* G *carries an abstract degree function* d *it admits a faithful representation* $\phi : G \to FGL(V)$ *for some vector space* V *over a (skew) field* K *such that the degree function of* FGL(V) *restricts to* d *on the image* G^ϕ *of* G.

Proof. Let $\mathfrak{F}$ denote the set of all finitely generated subgroups of G. For $F \in \mathfrak{F}$ choose a faithful representation ϕ_F of F into the group $GL(d(F), K_F)$ for some (skew) field K_F according to the definition of the abstract degree function d. For every $F \in \mathfrak{F}$ put

$$P_F := \{X \in \mathfrak{F};\ F \subseteq X\}.$$

This family of subsets defines a filter on $\mathfrak{F}$; let $\mathfrak{U}$ be any ultrafilter refining this filter. Then there is a natural embedding ϕ of the group G into the ultraproduct

$$\bar{G} = \prod_{F \in \mathfrak{F}} F/\mathfrak{U}.$$

This ultraproduct is a group of automorphisms of the vector space

$$\bar{V} = \prod_{F \in \mathfrak{F}} V_F/\mathfrak{U}$$

over the (skew) field $\bar{K} = \prod_{F \in \mathfrak{F}} K_F/\mathfrak{U}$. In fact, the image of G^{ϕ} in $\bar{G}$ is contained in $FGL(\bar{V})$, and for every finitely generated subgroup F of G the degree function of FGL(V) takes the value d(F) on the corresponding subgroup F^{ϕ} of $\bar{G}$.

It should be pointed out that the vector space $\bar{V}$ constructed in this proof has in general a much higher dimension than is necessary to obtain such a representation ϕ.

On the class $\mathbf{C}$ of all pairs (G,d), G a group and d an abstract degree function on G, one may refine the natural order relation of containment by setting

$(G,d_G) \leq (H,d_H)$ if and only if $G \subseteq H$ and d_G and d_H coincide on the finitely generated subgroups of G.

If I is a linearly ordered index set and $\tau = \{(G_i,d_i);\ i \in I\}$ is an ascending sequence of groups with degree functions in this order on $\mathbf{C}$, i.e. $(G_i,d_i) \leq (G_j, d_j)$ if and only if $i \leq j$, then, obviously, the group $T := \bigcup_{i \in I} G_i$ admits the degree function

$$d := \bigcup_{i \in I} d_i.$$

Thus, the group T is finitary linear; the class $\mathbf{C}$ is inductive for this order relation.

2. Existentially closed finitary linear groups

Let $\mathbf{C}$ be a class of structures with a first order language $\mathbf{L}$. The structure $E \in \mathbf{C}$ is *existentially closed in* $\mathbf{C}$ if for every structure $S \in \mathbf{C}$ with $E \subseteq S$ every finite set of (existential) equations and inequalities in $\mathbf{L}$ with coefficients in E which admits solutions in S also admits solutions in E. The existentially closed structures in $\mathbf{C}$ may be viewed as substitutes for maximal structures in $\mathbf{C}$ which rarely exist; they are smooth and big objects in $\mathbf{C}$. In general, however, there do not exist any existentially closed structures in the class $\mathbf{C}$ either. But if the class $\mathbf{C}$ is inductive, i.e. if the union of any ascending sequence of structures in $\mathbf{C}$ still belongs to $\mathbf{C}$, every structure $S \in \mathbf{C}$ is contained in some $E \in \mathbf{C}$ which is existentially closed in $\mathbf{C}$. (Essentially, this statement goes back to Steinitz [9], see also [5, p.19].)

The notion of structures existentially closed in $\mathbf{C}$ is a vast generalisation of that of algebraically closed fields in the class of fields. So one may wonder whether the existentially closed structures in $\mathbf{C}$ play a role in the theory of $\mathbf{C}$ comparable to that played by the algebraically closed fields in the theory of fields. For this reason there is some interest in studying the internal constitution and behaviour of structures E existentially closed in some particular class $\mathbf{C}$. For certain classes $\mathbf{C}$ of groups such a study has been done.

Here we shall consider the class $\mathbf{C}$ of finitary linear groups with degree function.

By the remark following Theorem 1 the class $\mathbf{C}$ of all finitary linear groups with degree function over (skew) fields is inductive. Let (E,d) be any existentially closed structure in this class $\mathbf{C}$. By Theorem 1 (E,d) has a realisation as a subgroup of FGL(V) for some vector space V over some (skew) field K which may be assumed existentially closed in the class of all (skew) fields. We shall study the structure of (E,d) and its relation to V and K.

Fact 1. *For every finite group* F *and every natural number* n *which is* F-*admissible over* K *there is a subgroup* F_n *of* E *with*

$$F \cong F_n \text{ and } d(F_n) = n.$$

Proof. Consider (E,d) realised as a subgroup of FGL(V). This is in turn a subgroup of FGL(U ⊕ V) where U is an n-dimensional vector space over K; the degree function d' of FGL(U ⊕ V) extends that of FGL(V) and hence d. Let the group F act admissibly on U and trivially on V, then F is a subgroup of

$$FGL(U \oplus V)$$

with d'(F) = n. Now, every finite group is completely described (up to isomorphism) by its multiplication table. So F solves the corresponding finite set of equations and inequalities (with coefficients $l \in E$) together with the equation d'(F) = n. Since (E,d) is existentially closed in $\mathbf{C}$ there must be a subgroup $F_n \subseteq E$ satisfying this very set of equations and inequalities:

$$F_n \cong F \text{ and } d'(F) = d(F) = n.$$

This fact yields that the function d is unbounded on the set of finitely generated subgroups of E. Consequently, the vector space V is infinite-dimensional.

Fact 2. *The characteristic* c *of the (skew) field* K *is uniquely determined by* (E,d). *It is the prime* p *precisely if there is a finite subgroup* F *in* E *with*

$$F \cong SL(2,p^{\alpha}),\ \alpha \geq 3,$$

and d(F) = 2.

Proof. If K has characteristic c = p then such a subgroup F exists for every $\alpha \in \mathbb{N}$ in FGL(V) since K is an existentially closed (skew) field. If, however, $c \neq p$ then the elementary abelian group of order p^{α}, the Sylow p-subgroup of F, does not admit a faithful representation of degree 2 over K.

Fact 3. *For the finitely generated subgroups* A, B *of* E *with*

$$d(A) \leq d(B) \leq n$$

there exists an element $g \in E$ *with* $d(\langle A,B^g\rangle) \leq n$.

Proof. Let the action of A on V be concentrated in the subspace U_A of V and that of B in U_B with $\dim_K U_A \leq \dim_K U_B \leq n$. Then there is an element $h \in FGL(V)$ with

$$U_A \subseteq (U_B)^h = U_{(B^h)}.$$

So the equation

$$\exists\, x \text{ with } d(\langle A,B^x\rangle) \leq n$$

admits a solution in FGL(V). Since (E,d) is existentially closed in $\mathbf{C}$ there must be such an element g in E.

Fact 4. *Let* F *be a finitely generated subgroup of* E *such that*

$$V = [F,V] \oplus C_V(F), \;\; \dim_K[F,V] = n = d(F).$$

The subgroup

$$R = \langle H \subseteq E;\; H \textit{ finitely generated with } d(\langle F,H\rangle) = n\rangle$$

of E *is existentially closed in the class* $\mathcal{L}_n(c)$ *of linear groups of degree* n *over (skew) fields of characteristic* c.

Proof. Observe that the action of R on V is concentrated in the subspace

$$U := [F,V]$$

and trivial on $C_V(R)$. Assume that the finite set Σ of equations and inequalities with coefficients in R admits solutions in a group T containing R which is linear of degree n over some (skew) field K' of characteristic c. Since the class of all (skew) fields has the amalgamation property [2], one may assume $K \subseteq K'$. Thus T acts faithfully on the vector space $U' := K' \otimes U$ over K' extending the action of R on U. Letting T act trivially on the space

$$K' \underset{K}{\otimes} C_V(F),$$

we embed T into FGL(V') with

$$V' := \underset{K}{\otimes} V$$

such that the subspace U' is admissible. If the finitely generated subgroup T_1 of T contains F, the coefficients of Σ and a solution set for Σ then one has for the degree function d' of FGL(V') that $d'(T_1) = n$. The function d' extends the degree function of FGL(V) and so d. Thus T_1 shows that the set Σ even admits solutions over FGL(V). Since (E,d) is existentially closed in $\mathbf{C}$ there must exist a finitely generated subgroup $T_2 \subseteq E$ containing F and a solution set for Σ. This subgroup T_2 must lie in R. So R is existentially closed in the class $\mathcal{L}_n(c)$.

Observe that the conditions on F in Fact 4 are met by any finite non-abelian simple subgroup of E for which the action on V is concentrated in an F-irreducible subspace. Since by Fact 1 there are such subgroups F with d(F) arbitrarily large, the subgroups R constructed in Fact 4 form a local system of E. The structure of the groups R was studied at least for groups existentially closed in the class of linear groups of degree n over commutative fields by Mez [7]. The argument and main result go over to the general case:

Fact 5. *If* L *is existentially closed in the class* $\mathfrak{L}_n$ *of linear groups of degree* $n \geq 3$ *over (skew) fields, then there is an existentially closed (skew) field* $\bar{K}$ *such that the commutator subgroup* L' *of* L *is isomorphic to*

$$SL(n,\bar{K}) = (GL(n,\bar{K}))'.$$

We are now in position to describe the structure of (E,d). This will extend (and use) the results of Mez and almost characterise the (E,d) existentially closed in $\mathfrak{C}$.

Theorem 2. *If* (E,d) *is existentially closed in* $\mathfrak{C}$, *then the centre of* E *is trivial and the commutator subgroup* E' *of* E *is simple. Further, there exists an existentially closed (skew) field* $\bar{K}$ *such that* E' *has a local system of stable special linear groups over* $\bar{K}$.

Proof. The finitely generated central subgroup Z of E determines the number d(Z). By Fact 1 there is a finitely generated subgroup F of E with d(F) > d(Z); by Facts 3 and 4 we may assume that $Z \subseteq F$, that the action of F on V is concentrated in [F,V], and that F acts irreducibly on [F,V]. By Schur's lemma the central subgroup Z of F acts fixed-point-freely (by multiplications) on [F,V] if $Z \neq \langle 1 \rangle$. On the other hand, the inequality

$$d(Z) < d(F) = \dim_K[F,V]$$

implies that Z has non-trivial fixed-points in [F,V]. This contradiction forces $Z = \langle 1 \rangle$. In E there is an ascending sequence $\{F_i\}$ of finitely generated subgroups F_i with $d(F_i) = n_i$ such that the action of F_i on V is concentrated in the subspace $[F_i,V] := U_i$ and irreducible in U_i. The subspaces $\{U_i\}$ form an ascending sequence of subspaces of V, the subgroup $\langle F_i; i \in \mathbb{N} \rangle$ of E acts stably and irreducibly on the countable-dimensional subspace $U := \bigcup_{i\in\mathbf{N}} U_i$ of V. As in Fact 4 consider the subgroups R_i of E defined by

$$R_i := \langle g \in E;\ d(\langle F_i, g\rangle) = n_i\rangle.$$

One has $R_i \subsetneq R_{i+1}$, and the group $R = \bigcup_{i\in\mathbf{N}} R_i$ acts stably and irreducibly on the subspace U of V. By Fact 4 each R_i is existentially closed in $\mathcal{L}_{n_i}(c)$, the class of all linear groups of degree n_i over (skew) fields of characteristic c. By Fact 5 there is an existentially closed (skew) field K_i such that $R_i' \cong SL(n_i, K_i)$. Since the action of R_i' on V is concentrated on the subspace U_i, one has that every element of R_{i+1}' leaving U_i invariant and acting trivially on $C_{U_{i+1}}(F_i)$ belongs to R_i. Thus $K_i = K_{i+1}$.

There exists an existentially closed (skew) field $\bar{K}$ $(= K_i) \subseteq K$ *so that*

$$R_i' \cong SL(n_i, \bar{K}) \textit{ for every } i \in \mathbf{N}.$$

So the group R' is the stable special linear group over $\bar{K}$; it is simple. By Fact 3 every finitely generated subgroup of E may be conjugated into one of the subgroups R_i by some element of E, so the commutator subgroup E' is simple. If $\{F_{1,i}\}$ and $\{F_{2,i}\}$ are two such ascending sequences of finitely generated subgroups of E which determine subgroups $R_{1,i}$ and $R_{2,i}$ and R_1 and R_2 in E, then the subgroup $\langle F_{1,i}, F_{2,i}\rangle$ is contained in a finitely generated subgroup $F_{3,i}$ such that the action of $F_{3,i}$ in V is concentrated in $[F_{3,i}, V]$ and irreducible. Choose $F_{3,i+1}$ to contain $F_{3,i}$. Then to the sequence $\{F_{3,i}\}$ belongs the ascending sequence $\{R_{3,i}\}$ with

$$R_3 = \bigcup_{i\in\mathbf{N}} R_{3,i}.$$

Clearly, the group $R_3' \supseteq R_1', R_2'$. So the subgroups of type R' form a local system of E'.

Remark. If the skew field $\bar{K}$ is existentially closed, the group $GL(n,\bar{K})$ equals $SL(n,\bar{K})$, since the multiplicative group $\bar{K}^*$ coincides with its commutator subgroup in this case. Since every element of $\bar{K}^*$ is a product of commutators, this is also true in R, and so E is a simple group.

Examples of pairs (E,d) which are existentially closed in $\mathbf{C}$ abound: Let $\bar{K}$ be any existentially closed (skew) field and V any infinite-dimensional vector space over $\bar{K}$. For E choose any subgroup of FGL(V) having a local system of subgroups (of type R) which are stable general linear groups over $\bar{K}$ (for countable-dimensional subspaces of V). If d is the restriction of the degree function of FGL(V) to E then the pair (E,d) is existentially closed in $\mathbf{C}$.

References

1. H Bass, K-theory and stable algebra, *Publ. I.H.E.S.* **22** (1964), 5-60.
2. P M Cohn, *Skew field constructions* (London Mathematical Society Lecture Note Series **27**, Cambridge University Press 1977).
3. J I Hall, Infinite alternating groups as finitary linear transformations, *J. Algebra* **119** (1988), 337-359.
4. J I Hall, Finitary linear transformation groups and elements of finite local degree, *Arch. Math.* **50** (1988), 315-318.
5. J Hirschfeld & W H Wheeler, *Forcing, arithmetic, division rings* (Lecture Notes in Mathematics **459**, Springer-Verlag 1975).
6. O H Kegel & B A F Wehrfritz, *Locally finite groups* (North-Holland Publishing Co. 1973).
7. H-C Mez, Existentially closed linear groups, *J. Algebra* **76** (1982), 84-98.
8. R E Phillips, The structure of groups of finitary transformations, *J. Algebra* **119** (1988), 400-448.
9. E Steinitz, Algebraische Theorie der Körper, *(Crelle's) J. Reine Angew. Math.* **137** (1910), 167-309.

PERMUTABILITY AND SUBNORMALITY OF SUBGROUPS

RUDOLF R MAIER

UnB-Brasilia, 70.910 Brasilia-DF, Brazil

The main purpose of this survey article is the presentation of a result of the author's paper: *Zur Vertauschbarkeit und Subnormalität von Untergruppen*, which appeared in *Archiv der Mathematik* **53** (1989),110-120.

The result was obtained while the author was visiting the University of Tübingen during a sabbatical semester in 1987.

For common concepts and notation which we shall use in this report, see introductory books in finite group theory, for example [G].

A. Introduction to the problem

If G is a group and if A,B are subgroups of G, the subgroup $\langle A,B\rangle$ of G generated by $A \cup B$ is of interest. Due to the fact that a group is a noncommutative algebraic structure, it is impossible in general to predict some special property of $\langle A,B\rangle$ only by the knowledge of the internal properties of the generating subgroups A and B. For example, two cyclic subgroups may generate a highly complicated simple group.

To be able to control the properties of the group $\langle A,B\rangle$ by those of A and B, the generation of $\langle A,B\rangle$ must happen in a special way. The most transparent case we have is when $\langle A,B\rangle$ coincides with the product set $AB = \{ab \mid a \in A, b \in B\}$. It is well known that this holds if and only if AB = BA. Two subgroups A and B of a group G which have this property, are called *permutable*. A sufficient condition for the permutability of A and B is that A normalizes B (that is, $a^{-1}ba \in B$ for all $a \in A$, $b \in B$) or vice versa. Particularly, if X is a normal subgroup of G ($X \trianglelefteq G$), we have $AX = XA = \langle A,X\rangle$ for every subgroup $A \leq G$.

In 1939, Ore [9] called a subgroup $X \leq G$ a *quasi-normal subgroup* of G (X qn G) if and only if AX = XA holds for all $A \leq G$. He proved that, in finite groups (and we shall restrict our considerations to this case), a quasi-normal subgroup is always a member of some composition series of G. The members of the composition series of G are those subgroups X of G for which there exists a chain of subgroups $X = X_0 \trianglelefteq X_1 \trianglelefteq \ldots \trianglelefteq X_{r-1} \trianglelefteq X_r = G$. The investigation of these subgroups, which received later on the name of *subnormal subgroups*, was initiated also in 1939 by Wielandt [10]. We write X sn G if X is a subnormal subgroup of G.

Questions on the relationship between permutability and subnormality of subgroups awakened interest. In 1962, Itô and Szép [3] obtained an interesting result which showed that the "difference" between normality and quasi-normality in general is small. Besides a quasi-normal subgroup being subnormal, we have: if X qn G, then the quotient group X/X_G is nilpotent, that is, X/X_G is contained in the Fitting subgroup $F(G/X_G)$ of G/X_G. Here X_G denotes the intersection of all conjugates $X^g = g^{-1}Xg$ of X with $g \in G$. The Fitting subgroup of a finite group is its largest nilpotent (sub-)normal subgroup.

On the other hand, there exist examples of quasi-normal subgroups which are not normal. The complete classification of the minimal groups which contain quasi-normal non-normal subgroups is given in [6], see also [5].

Still in 1962/63, Deskins [1] and Kegel [4] proved that, to ensure Itô and Szép's statement, the hypothesis of quasi-normality may be weakened: if X is a subgroup of G which permutes with all Sylow subgroups of G, we again have X sn G and $X/X_G \leq F(G/X_G)$. With respect to quasi-normal subgroups one can state essentially more: X qn G implies that X/X_G is contained in the hypercentre of the group G/X_G. The *hypercentre* of a finite group is the term from which point on its upper central series becomes stationary. This result was discovered in 1973 in [7].

B. m-embedding of subgroups

Definition. We call a subgroup X of a finite group G an m-*embedded subgroup* of G (X m G) if AX = XA holds for all maximal subgroups A of G.

Clearly, every quasi-normal subgroup of G is an example of an m-embedded subgroup of G. If $\Delta(G)$ denotes the intersection of the non-normal maximal subgroups of G and if $X \leq \Delta(G)$, then X m G. Particularly, in a nilpotent group ($\Delta(G) = G$), every subgroup is m-embedded. The subgroup $\Delta(G)$ was introduced by Gaschütz in [2] who also proved its nilpotency. For $X \leq \Delta(G)$ we therefore also have X sn G. (Each of the following conditions is equivalent to the nilpotency of a finite group G:

(a) The maximal subgroups of G are normal in G.

(b) All subgroups of G are subnormal in G.)

These examples justify the question: is it true that an m-embedded subgroup of a group G is always a subnormal subgroup of G? The answer is no, and, as we shall see, the results are more interesting. One can state the following:

Theorem 1. *Let* G *be a group and* X m G. *Then the Fitting subgroup* F(X) *is a subnormal subgroup of* G.

In particular, if X m G and if X/X_G is nilpotent, then in fact X is a subnormal subgroup of G. This has as a consequence that *soluble* m-embedded subgroups behave *as if they were* subnormal. To formulate this more exactly, we introduce: in every group G let $F_0(G) = 1$ and, for $i \geq 1$, let $F_i(G)$ be the subgroup of G such that $F_i(G)/F_{i-1}(G)$ is the Fitting subgroup of $G/F_{i-1}(G)$. For $0 \leq r \in \mathbf{Z}$, G is usually called a *soluble group of Fitting length* $\leq r$, if $F_r(G) = G$. Let also $O_\pi(G)$ denote the largest normal π-subgroup of G if π is a set of prime numbers. We have:

Theorem 2. *Let* G *be a group,* X m G, $0 \leq r \in \mathbf{Z}$ *and let* π *be a set of prime numbers. Suppose that* X *is a soluble* π*-group of Fitting length* $\leq r$. *Then*

$$X \leq F_r(G) \cap O_\pi(G).$$

This implies:

Consequence 1. *Let* X *be a Hall subgroup of* G *(that is, the order of* X *and the index of* X *in* G *are coprime numbers). Suppose that* X *is a soluble subgroup. Then we have:*

$$X \trianglelefteq\trianglelefteq G \Leftrightarrow X \text{ m } G.$$

In spite of these results we shall see that every non-nilpotent group can be m-embedded into a larger group without being subnormal in it.

C. Basics of general representation theory

Semi-direct products and modules. First we want to note some facts about semi-direct products and modules.

Let X and V be two groups. Suppose there is defined an "action" of X on V. This means that, for every $x \in X$, we have an automorphism $v \mapsto v^x$ of V and the following rules are satisfied:

$$v^{xy} = (v^x)^y, \ v^1 = v$$

for every choice of the elements $v \in V$, $x,y \in X$; 1 is the unit element of X. In this situation, the set $G = \{(v,x) \mid v \in V, x \in X\}$ is a group under the multiplication

$$(v,x)\cdot(w,y) = (vw^{x^{-1}},xy) \quad ((v,x),(w,y) \in G).$$

The group G is called the *semi-direct product* of V by X (with respect to the given action), denoted by G = [V]X.

The semi-direct product G = [V]X contains a copy $\bar{V} = \{(v,1) \mid v \in V\}$ of V as a normal subgroup and a copy $\bar{X} = \{(1,x) \mid x \in X\}$ of X as a subgroup. We have $\bar{V} \cap \bar{X} = \{(1,1)\}$ and $G = \bar{V}\bar{X}$. The given action of X on V corresponds to the conjugation of the elements of $\bar{V}$ by the elements of $\bar{X}$:

$$(1,x)^{-1}(v,1)(1,x) = (v^x,1).$$

One also says that X *acts trivially* on V if $v^x = v$ for all $x \in X$, $v \in V$. This happens if and only if G = [V]X is the direct product $G = V \times X$.

Of particular interest is the case when the group V is an elementary abelian p-group for some prime p. Now V can be considered as a vector space of finite dimension over the prime field **Z**/(p). It is usual to write V additively in this case. An action of X on V is now a *linear representation* of X over **Z**/(p). We say that V is an X-*module* over **Z**/(p).

Let V be an X-module over the field **Z**/(p). We have the following standard concepts: an X-*submodule* W of V is a subspace of V such that

$$W^x = \{v^x \mid v \in W\} = W$$

for all $x \in X$. V is an *irreducible* X-module, if V is of dimension > 0 without X-submodules ≠ V and ≠ {0}. If W is an X-submodule of V, it is clear that W and the quotient space V/W are also X-modules. W is called a *maximal* X-submodule of V if V/W is an irreducible module. The *radical* rad V of V is the intersection of all the maximal X-submodules of V. A *composition series* of V as an X-module is a chain of X-submodules

$$V = V_0 > V_1 > \dots > V_{r-1} > V_r = \{0\}$$

of V such that V_i is a maximal X-submodule of V_{i-1} for i = 1,...,r. The *composition factors* in this series are the irreducible modules V_{i-1}/V_i (i = 1,...,r). The composition factors of V are the composition factors in all possible composition series of V.

Subnormality and m-embedding in the semi-direct product. With the concepts introduced above it is not too difficult to prove:

Theorem 3. *Let* X *be a group,* p *a prime number,* V *an elementary abelian* p-*group which is an* X-*module. Let* G = [V]X. *Then the following hold:*

(a) X sn G, *if and only if* X *acts trivially on every composition factor of* V.

(b) X m G, *if and only if* X *acts trivially on* V/rad V.

Therefore we have: the subgroup X of G = [V]X is m-embedded but not subnormal in G, if and only if X acts trivially on V/rad V but non-trivially in some composition factor of V.

What are the groups X like which admit such a module V over some **Z**/(p)? This is solved by:

Theorem 4. *Let* X *be a group,* p *a prime number. The following statements are equivalent:*

(a) X *is a* p-*nilpotent group (that is,* X *has a quotient group isomorphic to the* p-*Sylow subgroups of* X).

(b) *For every* X-*module* V *over* **Z**/(p) *we have: if* X *acts trivially on* V/rad V, *then* X *acts trivially on every composition factor of* V.

A group is nilpotent if and only if it is p-nilpotent for all prime numbers p. Therefore we have:

Consequence 2. *Let* X *be a non-nilpotent group. Then:*

(a) *There exists a prime number* p *and an* X-*module* V *over* **Z**/(p) *such that* X *acts trivially on* V/rad V *but non-trivially on some composition factor of* V.

(b) *In the group* G = [V]X, *the subgroup* X *is* m-*embedded but not subnormal.*

Together with the result of Theorem 1, we have the following characterization of finite nilpotent groups:

Theorem 5. *Let* X *be a group. The following statements are equivalent:*

(a) X *is a nilpotent group;*

(b) *every m-embedding of* X *into a bigger group is subnormal.*

An open question. Is it true that every m-embedded Hall subgroup X of G is necessarily a (sub-)normal subgroup of G? For soluble Hall subgroups this is true (Consequence 1). In the examples we treated, the subgroup X which is m-embedded but not subnormal, has prime power index for some prime divisor of the order of X. Therefore these examples don't give an answer to our question.

D. Concrete examples

We mention a series of matrix groups which contain m-embedded non-subnormal subgroups.

Let p be a prime number, n a natural number. In the general linear group GL(n+2,p), let G be the subgroup of all matrices of the form

$$\begin{pmatrix} 1 & b_2 & b_3 & \dots & b_{n+2} \\ 0 & 1 & a_3 & \dots & a_{n+2} \\ 0 & 0 & & & \\ \cdot & \cdot & & & \\ \cdot & \cdot & & Y & \\ \cdot & \cdot & & & \\ 0 & 0 & & & \end{pmatrix}$$

where a_i, b_i are elements of $\mathbf{Z}/(p)$ and Y is an irreducible subgroup $\neq 1$ of GL(n,p). G is the semi-direct product of the elementary abelian group V of all matrices of the form

$$\begin{pmatrix} 1 & b_2 & b_3 & \dots & b_{n+2} \\ 0 & 1 & 0 & \dots & 0 \\ 0 & 0 & 1 & \dots & 0 \\ \cdot & \cdot & & & \cdot \\ \cdot & \cdot & & & \cdot \\ \cdot & \cdot & & & \cdot \\ 0 & 0 & 0 & \dots & 1 \end{pmatrix}$$

by the subgroup X of all matrices of the form

$$\begin{pmatrix} 1 & 0 & 0 & \dots & 0 \\ 0 & 1 & a_3 & \dots & a_{n+2} \\ 0 & 0 & & & \\ \cdot & \cdot & & Y & \\ \cdot & \cdot & & & \\ \cdot & \cdot & & & \\ 0 & 0 & & & \end{pmatrix}.$$

Let W be the hyperplane $b_2 = 0$ of V. It is not difficult to check the following:

(1) W is a non-trivial irreducible X-submodule of V and X acts trivially on V/W.

(2) W is the only X-submodule $\neq \{0\}$ and $\neq$ V of V. Therefore we have W = rad V.

(3) X is an m-embedded but not subnormal subgroup of G.

References

[G] D Gorenstein, *Finite Groups* (Harper and Row, New York, Evanston, London).

1. W E Deskins, On quasinormal subgroups of Finite groups, *Math. Z.* **82** (1963), 125-132.

2. W Gaschütz, Über die Φ-Untergruppe endlicher Gruppen, *Math. Z.* **58** (1953), 160-170.

3. N Itô & J Szép, Über die Quasinormalteiler von endlichen Gruppen, *Acta Sci. Math.* **23** (1962), 168-170.

4. O H Kegel, Sylowgruppen und Subnormalteiler endlicher Gruppen, *Math. Z.* **78** (1962), 205-221.

5. R Maier, *Bedingungen für die Normalität von Subnormalteilern in endlichen Gruppen* (Doctoral Thesis, Tübingen 1969).

6. R Maier, Normality Conditions for quasinormal subgroups in finite groups, *Math. Z.* **123** (1971), 310-314.

7. R Maier & P Schmid, The embedding of quasinormal subgroups in finite groups, *Math. Z.* **131** (1973), 169-272.

8. R Maier, Zur Vertauschbarkeit und Subnormalität von Untergruppen, *Archiv der Math.* **53** (1989), 110-120.

9. O Ore, Contributions to the theory of groups of finite order, *Duke Math. J.* **5** (1939), 431-460.

10. H Wielandt, Eine Verallgemeinerung der invariantent Untergruppen, *Math. Z.* **45** (1939), 209-244.

SOME APPLICATIONS OF POWERFUL P-GROUPS

AVINOAM MANN

Hebrew University, Jerusalem, Israel

In my talk in the second "Groups - St Andrews", I outlined the answer to a question that was asked by Jim Wiegold in the first such conference. Another application of the same methods was reported in the same meeting by Alex Lubotzky (see Sections 6 and 3 below). Now, in the third "Groups - St Andrews", I would like to report on further applications. Some of these were, in their turn, inspired by Dan Segal's talk in the second conference, and include a solution to a problem he asked.

As the reader will see, the proofs (which are due to A Lubotzky, D Segal, and the author) apply a variety of techniques and results. In addition to the powerful p-groups mentioned in the title, main new ingredients are the application of p-adic Lie groups, and counting the number of generators of subgroups. These three themes are closely related.

Sections 2 to 4 contain applications to infinite groups, and section 5 to finite p-groups, while the last section, which also discusses finite p-groups, is discursive and speculative.

1. Powerful p-groups

Definition. A finite p-group G is *powerful*, if either p is odd and G^p contains G', or p=2 and G^4 contains G'.

(G^n and G' denote, respectively, the subgroup generated by nth powers and the commutator subgroup.)

We had several reasons for considering such groups. First, they occur quite often. E.g. if G is a regular p-group, then G^p is powerful. The same is true if G satisfies the conditions, weaker than regularity, that were studied in [M1]. A still more general example is given in Theorem 12 below. As another example, in [K1], B W King shows that subgroups of G^p are often powerful (this is generalized in [LM1.I, Proposition 1.13]). Next, it has been noticed by several people that powerful groups enjoy many interesting properties. Thus, in [Ho] and

[Mc] inequalities for numbers of generators are proved. Indeed, the thesis of B W King develops a theory of powerful p-groups (under a different name), but this remained unpublished (except for [K1]), and we knew about it only from a short abstract [K2]. I have seen a copy of this thesis only after [LM1] was already submitted ([LM1] contains many, but by no means all, of King's results). Finally, powerful groups figure in Lazard's theory of p-adic analytic groups, which is discussed in the next section.

The following are the two main properties of powerful p-groups, as far as the current applications are concerned. In these, we denote by d(G) the minimal number of generators of G, and the *rank* rk(G) is max d(H), as H ranges over all finitely generated subgroups of G (these notations and terminology are used for all groups, not necessarily p-groups or finite).

Theorem 1 ([LM1.I, Theorem 1.12]). *Let* G *be powerful, and let* $H \subseteq G$. *Then* $d(H) \leq d(G)$.

Theorem 2([LM1.I, Theorem 1.14]). *Let* G *be of rank* r. *Then* G *contains a normal powerful subgroup of index at most* $p^{r\log_2 r+r}$.

Of these, Theorem 2 is an immediate corollary of [LM1.I, Proposition 1.13], mentioned above. A shorter proof, yielding slightly less, is given in [M2] (when we refer below to the proof of Theorem 2, we usually mean this shorter proof). Theorem 1 was proved in [Mc] for a special case, and it was originally proved by noting that the lemmas of [Mc] hold for all powerful groups. The proof of [LM1] shows how to construct a set of generators for H.

Problem 1. Can the bound for the index in Theorem 2 be improved? Can we, perhaps, replace the logarithm by a constant?

Of the other interesting properties of powerful p-groups, I mention only the following, which will be referred to in the last section.

Theorem 3([LM1.I, Theorem 1.11]). *Let* G *be a powerful p-group generated by the elements* $a_1,\dots,a_d$. *Then* G *is the product of the cyclic subgroups* $\langle a_1\rangle,\dots,\langle a_d\rangle$. *Conversely, if a finite* p*-group,* p *odd, is the product of* d *cyclic subgroups, with* $d = d(G)$, *then* G *is powerful.*

2. Pro-p groups and p-adic Lie groups

A pro-p group is, by definition, an inverse (projective) limit of finite p-groups. As an inverse limit, it carries a natural topology, induced by the discrete topology on its factor p-groups. This topology is thus compact and totally disconnected. If G is a pro-p group, we say that a set X of elements generates G, if G is the minimal closed subgroup containing X. In particular, G' and G^n denote the closed subgroups generated by the commutators and nth powers, respectively. The notation d(G), rk(G), and the term *rank*, also refer to this notion of generation. With this notation, we define a *powerful pro-*p *group* as one that satisfies $G^p \supseteq G'$ ($G^4 \supseteq G'$). It is then easy to see, that G is a powerful pro-p group, if and only if it is an inverse limit of powerful finite p-groups.

A topological group is termed a p-*adic analytic group* (or a p-*adic Lie group*), if it fulfills the conditions in the definition of a Lie group, but with the underlying topological space being $\mathbf{Z}_p$, the ring of p-adic numbers, with its natural topology, rather than the real field. This is essentially an algebraic, not topological or analytic, concept. M Lazard [La] has characterized the abstract groups which can be given the structure of a p-adic analytic group (this structure, if it exists, is unique). Let us say that a group G is *virtually* of some property P, if it contains a subgroup of finite index having the property P (e.g. a finite group is virtually trivial). Then, a compact p-adic analytic group is virtually a pro-p group. A finitely generated pro-p group, in turn, is p-adic analytic if, and only if, it is virtually powerful. Theorems 1 and 2 now yield:

Theorem 4 ([LM1.II, Proposition 2.2]). *Let* G *be a finitely generated pro-*p *group. Then the following are equivalent:*

(a) G *is* p-*adic analytic.*

(b) *There is a number* r, *such that all open subgroups of* G *can be generated by* r *elements.*

(c) *There is a number* r, *such that all closed subgroups of* G *can be generated by* r *elements.*

Here we refer of course to the number of topological generators. Note that an open subgroup of a pro-p group is closed and of finite index. Theorem 4 can be used to give alternative proofs of results about p-adic analytic groups, e.g. a theorem of Serre, that the class of p-adic analytic groups is extension closed, or the fact that p-adic analytic groups satisfy the maximum condition for closed subgroups.

Problem 2. Is a pro-p group satisfying the maximum condition for closed subgroups p-adic analytic?

Problem 3. Let G be a pro-p group. Suppose there exists a number r, such that each open subgroup of G contains an open subgroup with r generators. Is G p-adic analytic? If the answer is "no", characterize such groups.

See [LM1.II, p.512], to see where this problem comes from. If we change "open" to "open normal", the answer is "yes".

Given any group G, the set of all finite p-factor groups of G forms an inverse system of groups in a natural way. We denote by $\hat{G}$ the inverse limit of this system, called the pro-p *completion* of G. There is a canonical homomorphism

$$G \to \hat{G},$$

which is an embedding precisely when G is a residually-p group. In the next two sections, we study infinite residually-p groups, whose pro-p completion is "well behaved". Specifically, we would like $\hat{G}$ to be p-adic analytic (for a previous application of this idea, see [Lu1]).

More details about p-adic analytic groups will be found in [DDMS].

3. Linear groups

As is well known, linear groups are much better behaved than infinite groups in general, and there are sharper tools for studying them. They appear naturally in our context in two ways: first, a finitely generated linear group of characteristic 0 is virtually a residually-p group, for almost all (i.e. all but finitely many) primes p, while if the characteristic is finite the group is virtually residually-p, where now p is the characteristic. Next, a p-adic analytic group G has associated with it a finite dimensional Lie algebra L (over the field $\mathbf{Q}_p$ of p-adic numbers). As L has a faithful finite dimensional representation (Ado's Theorem), it can be shown that so has G, so G is linear (over $\mathbf{Q}_p$). G also acts on L, yielding a not necessarily faithful representation, but which still suffices for many of the applications described below (the kernel of this action is the FC-centre of G - the set of elements with a centralizer of finite index).

The virtual p-residuality of linear groups can be refined by means of the following concept.

Definition [Lu2]. Let p be a prime. A p-*congruence structure* on a group G, is a descending chain $\{N_i\}$ of normal subgroups of G, with $N_0 = G$, satisfying the following:

1. $|G:N_i|$ is finite.

2. There exists a number r, such that all the factor groups N_i/N_j, for $1 \le i \le j$ (not G/N_j!) are finite p-groups with r generators.

3. $\cap N_i = 1$.

The following remarkable result is proved in [Lu2].

Theorem 5. *Let* G *be a finitely generated group. Then* G *is a linear group over a field of characteristic* 0, *if and only if* G *has a* p-*congruence structure for some prime* p, *in which case it has such a structure for almost all primes.*

Let us remind the reader, that a group is linear in some given degree and characteristic, if and only if all its finitely generated subgroups are linear in that degree and characteristic. The proof of Theorem 5 applies Theorem 4 in both directions. First, if G has a p-congruence structure, Theorem 4 shows that $\hat{N}$ is p-adic analytic, hence linear, and then G is also linear. On the other hand, if G is linear in characteristic 0, and finitely generated, it is linear over some finitely generated integral domain, and from this it can be shown that it is linear over $\mathbf{Z}_p$ for almost all p. But $GL_n(\mathbf{Z}_p)$ is naturally a compact p-adic analytic group, and so has a p-congruence structure.

Theorem 5 suggests several invariants associated with an infinite linear group G. First, we may consider the set of primes for which G has a p-congruence structure. Second, we consider the set of primes for which G has a p-congruence structure with the least possible value of r. This set is usually infinite, with an infinite complement [Lu2]. Then we may look at this minimal value for r. This number has some claim to be called "the dimension of G", and is related to the minimal degree of a faithful representation of G as a linear group. A clarification of this relationship may play a role in the solution of the next problem.

Problem 4. Characterize linear groups of a given degree.

Problem 5. Characterize linear groups of finite characteristic.

In practice, it does not seem that for a specific group it is easier to decide whether it has a p-congruence structure, than it is to decide directly whether it is linear. Rather, as we will see in the next section, Theorem 5 is exploited in [LM2], [LM3], [MS] to show that groups satisfying certain assumptions must be linear, and then we try to sort out the linear groups satisfying these assumptions. For this "sorting out" we have to know something about linear groups. An extremely useful fact about linear groups is the "Tits' alternative" [T]. This states that if G is linear group which does not contain a non-abelian free group, then G is soluble-by-locally-finite, and if either G is finitely generated, or the relevant characteristic is 0, G is soluble-by-finite.

If our assumptions about G concern only subgroups of finite index, as seems natural in view of Theorems 4 and 5, the Tits' alternative seems to be inapplicable. A substitute is provided by the following result, implicit in [LM3].

Theorem 6 (Lubotzky's alternative). *Let* G *be a finitely generated group, linear in characteristic* 0. *Then either* G *is virtually soluble, or* G *contains a finite index subgroup* H, *and there exist a finite set of primes* Π, *and a non-trivial semi-simple connected algebraic group* **X** *(defined over* **Q**), *such that* H *has* **X**(**Z**/n**Z**) *as a factor group, for all* Π'*-numbers* n.

This theorem provides us with a large supply of finite index subgroups of G. We sketch briefly the proof. We may assume that G is a subgroup of GL(n,C). Then G is residually a subgroup of GL(n, $\bar{\mathbf{Q}}$). If all these $\bar{\mathbf{Q}}$-linear images are soluble, then, by a result of Zassenhaus, they are soluble of derived length bounded by a function of n, so G itself is soluble of that length. A less well known result of Platonov combines Zassenhaus' theorem with Jordan's theorem about the existence of abelian subgroups in finite linear groups, to show that there exist two functions of n, say f(n) and g(n), such that every virtually soluble subgroup of GL(n,C) has a soluble subgroup of index at most f(n) and length at most g(n) (see [We, 10.10]). Then the previous argument shows, that if G is not virtually soluble, we may assume that it is linear over **Q**. Then the identity component of the Zariski closure of G is not soluble, so it has a semi-simple (connected) image **X**. A slight change of groups allows us to assume even that **X** is simply connected, and then one is in a position to apply the deep "Strong Approximation Theorem", due to Nori and Matthews-Vaserstein-Weisfeller, which relates the Zariski and "pro-finite" closures of G [N], to end the proof.

4. Residually finite groups

The following result [LM2] is typical of our applications of the previous concepts and results to residually finite groups.

Theorem 7. *A residually finite group of finite rank is virtually locally soluble.*

Soluble groups of finite rank are reasonably well understood. For example a finitely generated such group is a so-called minimax group (see [R, 10.38]) - a fact that is used in the proof of Theorem 7 - and there are also results for the infinitely generated case. Thus this theorem provides a fair amount of information.

The proof of this result is described by the following diagram, in which the arrows in the upper row denote a series of reductions, while in the second row we describe the techniques which enable us to effect each reduction.

res fin --> res sol --> res-p --> pro-p --> p-adic Lie --> linear --> end

fin groups   inf sol groups   completion   lin groups

To be more specific, in the first reduction we use the Odd Order Theorem and a theorem of Tate, while in the second reduction we use the structure theorem for finitely generated soluble groups of finite rank, just mentioned. The passage from res-p to linear is clear, in view of the last two sections. Once we get to linear groups, the Tits' alternative (or, in this case, an earlier result of Platonov) finishes the proof.

The assumption of finite rank is one that imposes restrictions on all (finitely generated) subgroups of our group. As remarked already, it is natural to consider only subgroups of finite index. Moreoever, many of the arguments concern only finite factor groups of our groups. We thus introduce the following:

Definition. A group G has *upper rank* r, if all the finite factor groups of G have rank r.

In [S1], D Segal gives an example of a metabelian group of infinite rank and upper rank 2. This group is infinitely generated. Indeed we have [MS]:

Theorem 8. *A finitely generated residually finite group of finite upper rank is of finite rank, so its structure is given by Theorem 7.* (The possibility of changing finite rank to finite upper rank was first noticed by John Wilson.)

This is proved along the lines of the proof of Theorem 7, with the following differences. First, in the passage from res sol to res-p groups we apply more of the theory of infinite soluble groups, in particular Malcev's theorem that a linear soluble group is virtually triangularisable. This is applied to the action of G on its chief factors, and actually introduces one more step, res (nil by abelian), between the res sol and res-p steps. Second, in the linear group case we are not able to quote Tits' alternative. One way is to use Lubotzky's alternative, which, if our linear group is not virtually soluble, provides us with a finite factor group containing a 2-group of arbitrary large rank. In [MS] a more elementary approach is used, applying a result of Malcev stating that a finitely generated linear group is residually a finite linear group of the same degree. Finally, once we know that G is virtually soluble, and, being linear, it is also virtually res-p, the result of [S1] to be mentioned presently, shows (anticipating Theorem 9 (a)) that G is of finite rank.

We remark that the passage from res fin to res sol also employs only 2-groups. Thus we can see, that if in a residually finite group G all finite images have Sylow 2-subgroups of rank at most r (say), and G is locally of finite rank, then G is still virtually locally soluble, though it need not have finite rank.

We now come to the problem that motivated our consideration of residually finite groups. Let $f_G(n)$ denote the number of subgroups of G of index n. In [S1] D Segal posed the problem of characterizing the groups for which this function of n grows polynomially. We term such groups of *polynomial subgroup growth* (PSG). Segal proved that if G is finitely generated, soluble and residually nilpotent, then it has polynomial subgroup growth if and only if it has finite rank. This is not true if G is not finitely generated, as the example mentioned above shows. Indeed we have:

Theorem 9. (a) *A group of finite upper rank has polynomial subgroup growth (in particular, in such a group* $f_G(n)$ *is finite, even if* G *is not finitely generated).*

(b) *If* G *has polynomial subgroup growth, there exists a number* r, *such that all finite soluble images of* G *have rank* r.

This is proved essentially by counting, combined with some theory of finite soluble groups. In part (a), after reducing to soluble groups in the usual way, we use the existence and conjugacy of Sylow systems in these groups. In part (b) we need the fact that the order of a finite soluble group is bounded by a fixed power of the order of its Fitting subgroup ([Wo]; for the best known result of this type, see [Se]) and also an argument similar to the one that proves Theorem 2, and the fact that a "large" p-group has a lot of subgroups. This proof holds even if we assume polynomial growth only for the number of 2-subnormal subgroups.

I do not know if a finitely generated group of polynomial subgroup growth has a finite upper rank. Using the classification of the finite simple groups, one can show that the "upper composition factors" (i.e. composition factors of finite factor groups) of such a group are severely limited, but I do not know if there have to be only finitely many of them, which is necessary for finite upper rank. Thus, in the next result we are unable, so far, to perform the first reduction,

res fin --> res sol,

in the diagram above. The result (a combination of [LM3, MS]) is:

Theorem 10. *Let* G *be a finitely generated, residually (finite soluble) group of polynomial subgroup growth. Then* G *is a soluble minimax group.*

In the proof of this theorem, the passage from res sol to linear is the same as in Theorem 9. At the final stage, when we have a linear group, Lubotzky's alternative is combined with the Prime Number Theorem to give a finite image that not only contains a 2-group of large rank, but has itself an order which is bounded by some not too big function of the order of this 2-group. Then we count the subgroups of this 2-group to obtain a contradiction. This linear case is the heart of the proof. It is interesting to note that the result is true also for linear groups in finite characteristic, because they are virtually res-p, so we can apply Theorem 10 (or its res-p special case, which is somewhat shorter). In effect, we prove the result for finite characteristic by lifting, in a roundabout way, to characteristic 0.

The restrictions on upper composition factors of a PSG group mentioned above, which hold even if the group is infinitely generated, state that among these factors there occur only finitely many alternating groups, and that the groups of Lie type which occur are of bounded rank (here "rank" in its usual meaning in the Lie theory), and defined over fields whose dimension over the prime field is bounded. These restrictions and previous results combine to prove [MS]:

Theorem 11. *A finitely generated, residually finite group of polynomial subgroup growth, which satisfies the maximal condition for normal subgroups with residually finite factor groups, is virtually soluble of finite rank.*

There are examples of groups of polynomial subgroup growth, which have infinite upper rank and infinitely many upper composition factors. We can take, e.g., a direct product of infinitely many groups PSL(2,p_i), where the primes p_i are chosen to grow fast enough, and so that the orders of the various PSL(2,p_i)'s are nearly relatively prime. Of course these groups are not finitely generated. Similar examples are provided by the groups PSL(2,**R**), where **R** is **Z** localized at suitably chosen infinite set of primes. Thus we have even linear examples.

An apparently even more difficult problem is also posed in [S2]. Let G be a finitely generated group, and let

$$h_G(n) = |G:G^n|.$$

If G is soluble, this function of n is finite, so one can enquire about its behaviour. More generally, Zelmanov has recently announced the (positive) solution to the Restricted Burnside Problem, so we can define $h_G(n)$ differently, as the order of the maximal finite factor group of exponent n of G. The problem is then to characterize the groups for which this function grows polynomially. Let us call them groups of *polynomial power index growth.* They include all groups of finite upper rank (even if they are infinitely generated), but there are other examples. For example the positive solution to the congruence subgroup problem for SL(k,**Z**), k > 2 [Me2], shows that in this group G^n contains the kernel of the homomorphism onto SL(k,**Z**/n**Z**), this kernel having index less than n^k. So here we have a linear group of polynomial power index growth, which is nothing like virtually soluble. In [S2] it is proved, that a finitely generated soluble virtually residually nilpotent group has polynomial power index growth if, and only if, G has finite rank. Using a proof similar to the one of Theorem 2, it is not difficult to show that a finitely generated residually nilpotent group of polynomial power index growth has all of its finite nilpotent images of bounded rank, and is (hence) linear [MS]. Thus the following seems to be the natural next question.

Problem 6. Characterize linear groups of polynomial power index growth.

Let me remark, that while the function $f_F(n)$, where F is a free group, is an explicitly known upper bound for all the functions $f_G(n)$ (see [Ha, Theorem 7.2.9]; $f_F(n)$ is superexponential), we have no real idea what the similar function $h_F(n)$ looks like.

Let me make now some general vague remarks. For many years group theorists have been looking at infinite groups satisfying "finiteness conditions". However, constructions such as of Rips and Ol'shanskii [O], of infinite groups in which all proper subgroups have prime order, and other "monsters", show that even under very strong finiteness conditions, a group may behave wildly. It seems reasonable, then, to assume conditions which allow us to apply, rather than imitate, the theory of finite groups. A very successful illustration of this philosophy is the theory of locally finite groups, and residually finite groups seem to be the natural next case. The results of this section may, hopefully, be the first step in such a structure theory for residually finite groups, and the investigation of groups of polynomial power index growth may be a further step. This point of view suggested to me the next problem, which was open many years for general infinite groups, and has a positive answer for locally finite groups e.g. [R] (but a negative answer in general).

Problem 7. Does every infinite residually finite group contain an infinite abelian subgroup?

This is equivalent to:

Problem 7'. Does every infinite residually finite group contain a non-identity element with an infinite centralizer?

Naturally, we have to consider only torsion groups. J Wiegold has reduced Problem 7 to the following special case:

Problem 7*. Does every infinite residually-p group contain an infinite abelian subgroup?

In view of Zelmanov's theorem mentioned above, we may assume that the group has infinite exponent. We also take p to be odd, as all infinite 2-groups contain infinite abelian groups.

5. Moderate p-groups

In [M2], I discussed classes of p-groups defined by means of inequalities between the numbers of generators of their subgroups. Perhaps the most interesting among these classes is the class of *moderate* p-groups, where I call a p-group G *moderate*, if it satisfies

$$\text{If } H \subseteq K \subseteq G,\ \ |K:H| = p, \ \text{ and } \ K \text{ is not cyclic, then } d(H) - 1 < p(d(K)-1).$$

In other words, the Schreier inequality is always strict. The results outlined in [M2] imply then, that for H and K as above, we even have

$$d(H)-1 \leq (p-1)(d(K)-1).$$

The class of moderate groups includes all regular p-groups, and indeed the more general classes discussed in [M1]. It is the largest of the classes mentioned in [M2], and for a long time I wondered if it satisfies any structural constraints. This is answered by the following, which generalizes our first example of powerful p-groups:

Theorem 12. *Let* G *be a moderate* p-*group*, p *odd. Then* G^p *is powerful.*

This is proved by showing first:

Theorem 13. *Let* G *be a moderate* p-*group*, p *odd, and let* E *be an elementary abelian normal subgroup of* G. *Then* G^p *centralizes* E.

Theorem 12 follows by applying Theorem 13 to $G/(G^p)^p$. It is interesting to note that in a regular p-group, G^p centralizes all elements of order p, while under the more general assumptions of [M1], G^p still centralizes all normal subgroups of exponent p. We can also apply Theorem 13 to some subgroups of G, obtaining many powerful sections of G. These, in turn, allow us to apply results of P Hall, which guarantee that the power structure of G is, in some sense, "well behaved". These results of Philip Hall appear in an unpublished manuscript, containing results similar to, but more general than, the results of [M1], which it apparently predates (the results of [M1] are not general enough for the present application). They can also be applied to powerful p-groups (see [M2]). The author is extremely grateful to Norman Blackburn for providing him with a copy of this manuscript.

Theorem 14. *There exists a function* f(d,p), *such that all moderate* p-*groups* (p *odd) with* d *generators have rank at most* f(d,p).

This is obtained by combining Theorem 12 with Kostrikin's solution of the Restricted Burnside Problem for exponent p. This solution shows, that given d(G), the index $|G:G^p|$, and hence also $d(G^p)$, is bounded. Now apply Theorem 1.

If G is pro-p group, we may call it *moderate* if it satisfies the same inequality as before, for H and K of finite index. Theorem 13 shows then, that moderate pro-p groups are p-adic analytic. On the other hand, it is shown in [Lu3] (see also [I]), that if G is a pro-p group, in which the Schreier inequality is always an equality, then G is a free pro-p group. Thus free and p-adic analytic groups appear as the two extreme ends in the universe of pro-p groups.

We can also consider additive, rather than multiplicative, inequalities. Thus, let A_k be the class of p-groups such that, whenever H and K are as above, we have

$$d(H) \leq d(K)+k.$$

Then it can be shown that, if G is in A_k, and $k \leq p^n-2$, then G^p is powerful. It follows that the analogous result to Theorem 14 holds (here we do not need Kostrikin's Theorem). E.g. 2-generator groups in A_1 have rank at most 7 (I do not claim that this is best possible). Again, we can define similar classes of pro-p groups, and they will consist of p-adic analytic groups.

6. The number of relations of p-groups

The question of Wiegold, referred to in the opening sentence of this paper, asked for a bound for $t = d(M(G))$, where M(G) is the Schur multiplier of the p-group G, in terms of $r = rk(G)$. In [LM1] we gave the answer

$$t \leq r^2 \log_2 r$$

(for odd p; a slightly worse bound holds for p=2). The main point was, that if G is powerful, then $t \leq \binom{d+1}{2}$, and then we apply Theorem 2. If G is powerful, we also have $t \geq \binom{d}{2}$, and the question arises, can we narrow this interval in which t lies even further. The answer is "no". Indeed, it was during the 1985 "Groups - St Andrews" that I noticed that metacyclic groups with 2 relations provide me with some relevant examples (recall that t is the minimal number of relations needed to define G as a p-group). For p odd, the metacyclic p-groups are exactly the 2-generator powerful p-groups. To get better examples, I needed 3-generator 3-relation powerful p-groups. One of the organizers has kindly provided me with a copy of the paper [JR], which surveys all known finite groups with deficiency 0 (i.e. as many generators as relations). What I was looking for was rather special: most p-groups are not powerful, and very few of them have 3 generators and 3 relations. However, I needed only one example, so allowed myself to be hopeful. I was in for a surprise: all known p-groups of deficiency 0 are powerful!

Problem 8. Is this an accident?

It may well be. The first example of a 3-generator 3-relation group was given by Mennicke [Me1]. It is powerful; indeed the relations exhibit the commutators in terms of the pth powers, and it seems likely that they were chosen so, in order to make calculations feasible, and all later constructions have preserved this feature. A variation on Problem 8 was asked already by Wamsley, namely are all p-groups of deficiency 0 tricyclic (products of 3 cyclic groups)? (See [Wi, p.152], but right after that Wiegold poses a problem going in the other direction). Powerful 3-generator groups are tricyclic (Theorem 3), but powerful groups are more tractable than tricyclic ones. Problem 8 is probably very difficult. The following seems to be more modest, even though it may be asking for all p-groups of deficiency 0.

Problem 9. Determine the powerful p-groups of deficiency 0.

Let me end by pointing out the following corollary of the Golod-Shafarevitch inequality. Suppose that a p-group has d generators and r relations. Let H be a subgroup of index p^n, say. Then H has at most rp^n relations, so, by Golod-Shafarevitch, at most $2\sqrt{rp^n}$ generators. If d is close to r, and n is big, this may be a much better inequality than the one given by Schreier's formula. Thus it seems possible that a further study of groups satisfying sharp inequalities for the numbers of generators may be helpful for the investigation of groups with few relations.

Note added in proof

Since writing this paper, I have noticed that examples of 2-groups of deficiency zero that are not powerful were given already by one of the organizers, see E F Robertson, Canad. Math. Bull. **23** (1980), 313-316. More recently, 3-groups of this type were given by J Wiegold, Bull. Austral. Math. Soc. **40** (1989), 331-332.

Acknowledgement

The author is grateful to the Mathematical Institute of the University of Oxford, for its hospitality while this paper was being written, to the SERC for its support, and to Alex Lubotsky and Dan Segal for helpful comments on earlier drafts.

References

[DDMS] J D Dixon, M Desautoy, A Mann & D Segal, *p-adic analytic groups* (LMS Lecture notes series, in preparation).

[Ha] M Hall, *The theory of groups* (New York, MacMillan 1959).

[Ho] C R Hobby, A characteristic subgroup of a p-group, *Pacific J. Math.* **10** (1960), 853-858.

[I] I Ilani, Counting finite index subgroups and the P Hall enumeration principle, *Israel J. Math.* **68** (1989), 18-26.

[JR] D L Johnson & E F Robertson, *Finite groups of deficiency zero* (Homological group theory, Cambridge 1979), 275-289.

[K1] B W King, *Normal subgroups of groups of prime-power order* (Proc. 2nd Int. Conf. Theory of Groups, LNM **372**), 401-408.

[K2] B W King, Normal structure of p-groups, *Bull. Austral. Math. Soc.* **10** (1974), 317-318.

[La] M Lazard, Groupes analytiques p-adiques, *Publ. Math. IHES* **26** (1965), 389-603.

[Lu1] A Lubotzky, Group presentation, p-adic analytic groups and lattices in $SL_2(C)$, *Ann. of Math.* **118** (1983), 115-130.

[Lu2] A Lubotzky, A group theoretic characterization of linear groups, *J. Algebra* **113** (1988), 207-214.

[Lu3] A Lubotzky, Combinatorial group theory for pro-p groups, *J. Pure Appl. Algebra* **25** (1982), 311-325.

[LM1] A Lubotzky & A Mann, Powerful p-groups I & II, *J. Algebra* **105** (1987), 484-515.

[LM2] A Lubotzky & A Mann, Residually finite groups of finite rank, *Math. Proc. Cambridge Philos. Soc.* **106** (1989), 385-388.

[LM3] A Lubotzky & A Mann, On groups of polynomial subgroup growth, Invent. Math., to appear.

[M1] A Mann, The power structure of p-groups I, *J. Algebra* **42** (1976), 121-135.

[M2] A Mann, *Generators of p-groups* (Proc. Groups - St Andrews 1985, Cambridge 1986), 273-281.

[M3] A Mann, *The power structure of some p-groups* (Proc. Group Theory Conference, Bressanone 1989), to appear.

[MS] A Mann & D Segal, Uniform finiteness conditions in residually finite groups, *Proc. London Math. Soc.*, to appear.

[Mc] J J McCutcheon, A class of p-groups, *J. London Math. Soc.* (2) **5** (1972), 79-84.

[Me1] J L Mennicke, Einige endliche Gruppen mit drei Erzeugenden und drei relationen, *Arch. Math.* **10** (1959), 409-418.

[Me2] J L Mennicke, Finite factor groups of the unimodular group, *Ann. of Math.* **81** (1965), 31-37.

[N] M Nori, On subgroups of $GL_n(F_p)$, *Invent. Math.* **88** (1987), 257-275.

[O] A Yu Ol'shanskii, *On a geometric method in the combinatorial group theory* (Proc. Int. Congress Math. 1983, Warsaw 1984).

[R] D J S Robinson, *Finiteness conditions and generalized soluble groups* (Berlin, 1972).

[S1] D Segal, *Subgroups of finite index I* (Proc. Groups - St Andrews 1985, Cambridge 1986), 307-314.

[S2] D Segal, *Subgroups of finite index II* (Proc. Groups - St Andrews 1985, Cambridge 1986), 315-319.

[Se] Y Segev, On completely reducible solvable subgroups of GL(n,Δ), *Israel J. Math.* **51** (1985), 163-176.

[T] J Tits, Free subgroups in linear groups, *J. Algebra* **20** (1972), 250-270.

[W1] B A F Wehrfritz, *Infinite linear groups* (Berlin 1973).

[Wi] J Wiegold, *The Schur multiplier: an elementary approach* (Groups - St Andrews 1981, Cambridge 1982), 137-154.

[Wo] T R Wolf, Solvable and nilpotent subgroups of $GL(n,q^m)$, *Canad. J. Math.* **34** (1982), 1097-1111.

COMBINATORIAL ASPECTS OF FINITELY GENERATED VIRTUALLY FREE GROUPS

THOMAS MÜLLER

Universität Frankfurt am Main, West Germany

1. Introduction

Let G be a finitely generated (infinite) virtually free group and denote by m_G the least common multiple of the orders of the finite subgroups in G. The purpose of this note is to survey some recently obtained results centring around the arithmetic function $b_G : \mathbf{N} \to \mathbf{N}$,

$$b_G(\lambda) = \text{number of free subgroups of index } \lambda m_G \text{ in G},$$

which indicate a quite intimate relationship between combinatorial and structural properties of G. Proofs and more details will be given elsewhere. Our main results include an asymptotic formula for $b_G(\lambda)$ in terms of a presentation of G as a fundamental group of a finite graph of finite groups in the sense of Bass and Serre and an explicit estimate on the growth behaviour of the difference function $b_G(\lambda+1) - b_G(\lambda)$. Furthermore we consider the question as to what kind of information on the structure of G is contained in the number of free subgroups of finite index and as to what extent this combinatorial information determines the group G.

I would like to thank Robert Bieri and Heinrich Niederhausen for stimulating discussions during the preparation of this work.

2. Connection with Stallings' structure theorem

Our approach to b_G is to relate it to another arithmetic function $a_G : \mathbf{N} \to \mathbf{Q}$ which is certainly easier to compute. By a *torsion-free* G*-action* on a set X we mean a G-action on X which is free when restricted to finite subgroups. For a finite set X to admit a torsion-free G-action it is necessary and sufficient that $|X|$ be divisible by m_G. Now let μ be a positive integer. We define the positive rational number $a_G(\mu)$ by the condition that

$$(\mu m_G)!\, a_G(\mu) = \text{number of torsion-free G-actions on a set with } \mu m_G \text{ elements.}$$

The relationship between the two functions a_G and b_G is most conveniently described in terms of formal power series. With the convention $a_G(0) = 1$ introduce

$$\alpha_G(z) = \sum_{\mu=0}^{\infty} a_G(\mu) z^{\mu} \quad \text{and} \quad \beta_G(z) = \sum_{\lambda=0}^{\infty} b_G(\lambda+1) z^{\lambda}.$$

Then we have

$$\beta_G(z) = m_G \frac{d}{dz} (\ln \alpha_G(z)). \tag{1}$$

Formula (1) is contained in a more general counting principle which also incorporates M Hall's result on the number of subgroups of given finite index in a finitely generated free group, [H], as well as the generalizations of Hall's formula to (non-trivial) free products and free products with amalgamation obtained by I M S Dey and S G Ivanov respectively, cf. [D] and [I]. Comparing coefficients in (1) we find that the functions a_G and b_G are related by the transformation formula

$$\sum_{\mu=1}^{\lambda} a_G(\lambda-\mu)\, b_G(\mu) = m_G\, \lambda\, a_G(\lambda), \quad \lambda \geq 1. \tag{2}$$

Our next task then is to compute $a_G(\mu)$. For this we use Stallings' structure theorem, [St], which exhibits G as the fundamental group of a finite graph of groups $(G(-),Y)$ with finite vertex groups $G(v)$, cf. [KPS]. The fact that, conversely, the fundamental group of a finite graph of finite groups is always virtually free of finite rank is more elementary (see [Se, II.2.6, Prop. 11]).

Assuming such a decomposition

$$G \cong \pi_1(G(-),Y)$$

to be known for the group G we have a closed formula for $a_G(\mu)$ at hand. Writing V for the set of vertices and E for the set of (geometric) edges of Y we find that

$$a_G(\mu) = \frac{\prod_{e \in E} \left((\mu m_G/|G(e)|)!\; |G(e)|^{\mu m_G/|G(e)|}\right)}{\prod_{v \in V} \left((\mu m_G/|G(v)|)!\; |G(v)|^{\mu m_G/|G(v)|}\right)}, \quad \mu \geq 0. \tag{3}$$

3. The type $\tau(G)$

A first inspection of formula (3) reveals the noteworthy fact that the function a_G depends solely upon the graph Y and the order of the vertex- and edge- groups but not upon the internal structure of the G(v) and G(e). It is thus perfectly clear that the pair (m_G, b_G) does not in general determine the group G up to isomorphism. The question therefore arises: *what kind of information* on the structure of G is contained in the pair (m_G, b_G) and *to what extent* does this combinatorial information determine the group G?

Call two groups G and H *equivalent*, $G \sim H$, if

$$(m_G, b_G) = (m_H, b_H),$$

i.e. if G and H contain the same number of free subgroups for each given finite index. Let $(G(-),Y)$ be a finite graph of finite groups with fundamental group $\pi_1(G(-),Y) \cong G$ and for $j = 1,2,...,m_G$ put

$$\zeta_j = |\{e \in E \mid |G(e)| \mid j\}| - |\{v \in V \mid |G(v)| \mid j\}|.$$

Define the *type*

$$\tau = \tau(G(-),Y) \text{ of } (G(-),Y)$$

to be the tuple $(m_G; \zeta_1, \zeta_2, ..., \zeta_{m_G})$. By an easy calculation we derive from (3) the important relation

$$m_G \lambda \frac{a_G(\lambda)}{a_G(\lambda-1)} = (m_G \lambda)^{1+\zeta_{m_G}} \prod_{j<m_G} ((\lambda-1)m_G + j)^{\zeta_j}, \quad \lambda \geq 1 \tag{4}$$

a thorough discussion of which in conjunction with the transformation formula (1) leads to our first main result.

Theorem A. (a) *The type τ is an invariant of the group G i.e. independent of the decomposition of G in terms of a graph of groups*

$$(G(-),Y). \ \tau(G) = \tau(G(-),Y)$$

is called the type of G and we sometimes write $\zeta_j(G)$ for ζ_j.

(b) *For finitely generated virtually free groups G and H we have $G \sim H$ if and only if $\tau(G) = \tau(H)$.*

It follows, in particular, that the (rational) Euler characteristic of G is an invariant of the equivalence class,

$$\chi(G) = -\, m_G^{-1} \sum_j \zeta_j(G),$$

an observation which plays an essential role in the proof of the following finiteness result.

Theorem B. *An equivalence class of finitely generated virtually free groups contains only a finite number of non-isomorphic groups.*

We conclude this section by briefly mentioning another application of the techniques so far developed which although interesting in its own right will play no part in the further discussion. Equation (4) implies that the generating function $\alpha_G(z)$ associated with a finitely generated virtually free group G of type

$$\tau(G) = (m_G;\, \zeta_1, \ldots, \zeta_{m_G})$$

satisfies a homogeneous linear differential equation

$$A_0\, \alpha_G(z) + (A_1 z - m_G)\, \alpha_G'(z) + \sum_{\nu=2}^{\mu(G)} A_\nu\, z^\nu\, \alpha_G^{(\nu)}(z) = 0 \tag{5}$$

of order $\mu(G)$ with integral coefficients

$$A_\nu = \frac{1}{\nu!} \sum_{i=0}^{\nu} (-1)^{\nu-i} \binom{\nu}{i} m_G(i+1) \prod_{j=1}^{m_G} (i m_G + j)^{\zeta_j} \tag{6}$$

for $0 \leq \nu \leq \mu(G)$. Here $\mu(G)$ denotes the (free) rank of G defined in Section 5. Equation (5) is transformed via (1) into a differential equation of generalized Riccati-type for the series $\beta_G(z)$. Comparing coefficients we then deduce a recursion formula for the sequence $b_G(\lambda)$ which differs much in form from (2) and is particularly well adapted in pursuing number-theoretic aspects like for example divisibility properties of $b_G(\lambda)$, compare [M, §3]. The disadvantage of this approach to $b_G(\lambda)$ lies in the fact that the calculations involved in actually performing the transformation (1) become exceedingly complex if the order of the linear differential equation (5) is large ($\mu(G) \geq 10$ say).

4. Finite order graphs and their associated triangles

For reasons which will be explained in greater detail in Section 5 the discussion of analytical properties of the sequence $(b_G(\lambda))_{\lambda=1,2,\dots}$ renders it necessary to introduce a new quantity V_μ^λ obtained from the sequence $a_G(\mu)$ by an averaging process,

$$V_\mu^\lambda = \frac{a_G(\lambda)}{a_G(\mu)\, a_G(\lambda-\mu)}, \qquad 0 \le \mu \le \lambda.$$

If the group G is presented as a fundamental group of a finite graph (G(-),Y) of finite groups by (3) V_μ^λ takes on the form

$$V_\mu^\lambda = \frac{\prod_{e\in E} \binom{\lambda m_G/|G(e)|}{\mu m_G/|G(e)|}}{\prod_{v\in V} \binom{\lambda m_G/|G(v)|}{\mu m_G/|G(v)|}}. \tag{7}$$

"Forgetting" about the structure of the vertex- and edge groups in (G(-),Y) and keeping in mind only the order of the G(v) and G(e) leads to the concept of a *finite order graph*. More precisely an order graph $\underline{Y} = (n(-),Y)$ consists, by definition, of a connected graph Y together with functions $V \to N$ and $E \to N$ denoted by $v \mapsto n(v)$ and $e \mapsto n(e)$ respectively. We require $n(e) \mid n(v)$ whenever the vertex v is on the boundary of the edge e. The order graph $\underline{Y}$ is called *finite*, if Y is a finite graph. With a finite order graph $\underline{Y} = (n(-),Y)$ we associate its *triangle*

$$V_\mu^\lambda(\underline{Y}) = \frac{\prod_{e\in E} \binom{\lambda n/n(e)}{\mu n/n(e)}}{\prod_{v\in V} \binom{\lambda n/n(v)}{\mu n/n(v)}}, \qquad 0 \le \mu \le \lambda,$$

where n denotes the least common multiple of the vertex orders n(v). $V_\mu^\lambda(\underline{Y})$ is *symmetric*,

$$V_\mu^\lambda(\underline{Y}) = V_{\lambda-\mu}^\lambda(\underline{Y}),$$

and satisfies the boundary condition $V_0^\lambda(\underline{Y}) = 1$, $\lambda \ge 0$. We define the type $\tau(\underline{Y})$ by analogy with that of a graph of groups as $\tau(\underline{Y}) = (n; \zeta_1,\dots,\zeta_n)$ where

$$\zeta_j = |\{e \in E \mid n(e) \mid j\}| - |\{v \in V \mid n(v) \mid j\}|$$

and the rank $\mu(\underline{Y})$ of $\underline{Y}$ by the equation

$$\mu(\underline{Y}) = 1 + \Sigma \zeta_j \; (\mu(\underline{Y})$$

corresponds to the notion of "free rank" for finitely generated virtually free groups, cf. Section 5).

Finite order graphs and their associated triangles will be investigated more systematically in a subsequent paper. Here we confine ourselves to the remark that, philosophically speaking, analytical properties of the function b_G like those discussed in Section 5 are offshoots of the fact that under the assumption $\mu(\underline{Y}) \geq 2$ the triangle $V^{\lambda}_{\mu}(\underline{Y})$ stabilizes to an object with rather nice combinatorial properties ("nice" being measured relative to Pascal's triangle) *independent of the particular order graph* $\underline{Y}$ *under consideration.*

The following somewhat technical result should be seen as an illustration of this rather vague statement.

Theorem C. *Let* $\underline{Y}$ *be a finite order graph. Then*

(a) $V^{\lambda}_{\mu}(\underline{Y}) \geq \binom{\lambda}{\mu}^{\mu(\underline{Y})-1}, \quad 0 \leq \mu \leq \lambda.$

(b) *The triangle* $(V^{\lambda}_{\mu}(\underline{Y}))_{0\leq\mu\leq\lambda}$ *satisfies the unimodal property*

$$V^{\lambda}_{\mu-1}(\underline{Y}) < V^{\lambda}_{\mu}(\underline{Y}) \quad \textit{for} \quad 0 < \mu < \frac{\lambda+1}{2} \quad \textit{and} \quad \lambda \geq 2$$

provided that $\mu(\underline{Y}) \geq 2$.

An important consequence is :

Corollary C'. *For a finite order graph* $\underline{Y}$ *of rank* $\mu(\underline{Y}) \geq 2$ *we have*

$$A_\lambda := \sum_{\mu=1}^{\lambda-1} (V^{\lambda}_{\mu}(\underline{Y}))^{-1} = O(\lambda^{-s}),$$

where $s = \mu(\underline{Y}) - 1$, *in particular*

$$A_\lambda \to 0 \textit{ as } \lambda \to \infty.$$

5. Growth behaviour and asymptotics of the function b_G

The original motivation for the research reported here was to investigate the growth behaviour of the sequence $b_G(\lambda)$ associated with a finitely generated virtually free group G and in particular to supply a proof of the following conjecture which is now a theorem.

Theorem. *The sequence* $b_G(\lambda)$ *is strictly increasing provided that the group* G *contains a non-abelian free subgroup.*

It turns out that problems of this kind are of a rather subtle nature. The main reason for this lies in the fact that the transformation given by formula (1) is not well behaved when one tries to transmit *concrete properties* from a_G to the function b_G (as opposed to transforming the information involved in either of these functions as a whole). The situation becomes manageable through the introduction of the triangle V_μ^λ defined in the previous section. Our aim here is to describe two general results concerning the growth behaviour and the asymptotics of the function b_G and to explain their connection with Theorem C.

For a moment adopt the following more formal point of view. Consider two sequences

$$1 = a_0, a_1, a_2, \ldots \text{ and } b_1, b_2, b_3, \ldots$$

of real numbers satisfying a transformation formula

$$\sum_{\mu=1}^{\lambda} a_{\lambda-\mu}\, b_\mu = c\,\lambda\, a_\lambda, \qquad \lambda \geq 1 \tag{8}$$

with some constant $c > 0$. We require $b_j \geq 0$ and $a_i > 0$. Define the *triangle* $\Delta = (V_\mu^\lambda)_{0\leq\mu\leq\lambda}$ *associated with the transformation* (8) by

$$V_\mu^\lambda = \frac{a_\lambda}{a_\mu\, a_{\lambda-\mu}}, \qquad 0 \leq \mu \leq \lambda,$$

and for $\lambda \geq 1$ put

$$A_\lambda = \sum_{\mu=1}^{\lambda-1} (V_\mu^\lambda)^{-1}.$$

Our main tool for extracting information on the growth behaviour of the sequence b_λ is the following result.

Lemma. *Suppose that*

(i) $V_\mu^\lambda \geq \binom{\lambda}{\mu}^s$ *for* $0 \leq \mu \leq \lambda$ *and some fixed* $s \geq 1$

and

(ii) $a_1 \geq 1/2$.

Then the sequence b_λ *is strictly increasing and we have for* $\lambda \geq 1$

$$b_{\lambda+1} - b_\lambda > c\, a_1^\lambda\, (\lambda!)^s\, q(\lambda),$$

where $q(\lambda)$ *is the polynomial of degree* $s+1$ *given explicitly by*

$$q(\lambda) = a_1(\lambda+1)^{s+1} - 2a_1(\lambda+1) - \lambda.$$

As concerns the asymptotic behaviour of b_λ the following observation is due to E M Wright ([W, Theorem 3]) and independently to M Newman, [N].

If in the context of (8) $A_\lambda \to 0$ *as* $\lambda \to \infty$ *then* $b_\lambda \sim c\, \lambda\, a_\lambda$. (9)

In the case that $A_\lambda = O(\lambda^{-s})$ with some $s > 0$ the proof in fact shows that

$$1 = b_\lambda / c\lambda a_\lambda + O(\lambda^{-s}).$$

In view of the assumption (ii) in order to apply our lemma to the transformation (2) we need the following result which is also of some independent interest.

Proposition. *For a finitely generated infinite virtually free group* G *we always have* $a_G(1) \geq 1/2$ *with equality occurring if and only if either* G *is a generalized free product* $G = G_1 \underset{S}{*} G_2$ *with both* G_1 *and* G_2 *finite and*

$$[G_1{:}S] = 2 = [G_2{:}S]$$

or G *is a free product of finitely many (but at least three) copies of the cyclic group of order* 2.

We are now in a position to explain our main results. Define the *(free) rank* $\mu(G)$ of a finitely generated virtually free group G to be the rank of a free subgroup of index m_G in G. The existence of such a subgroup follows for example from (3) and (2). Observe that $\mu(G)$ is connected with the Euler characteristic of G via

$$\mu(G) + m_G\, \chi(G) = 1$$

which shows in particular that $\mu(G)$ is well defined. The arithmetic functions a_G and b_G associated with G are related by the transformation formula (8) with $c = m_G$. Choose a finite graph (G(-),Y) of finite groups representing G,

$$\pi_1(G(-),Y) \cong G,$$

and denote by $\underline{Y}$ the finite order graph canonically obtained from (G(-),Y). By (7) the triangle $(V_\mu^\lambda(\underline{Y}))_{0\leq\mu\leq\lambda}$ associated with $\underline{Y}$ coincides with the triangle Δ defined in connection with (8), and (by definition of the type of G) we have $\tau(G) = \tau(\underline{Y})$ implying $\mu(G) = \mu(\underline{Y})$. Now assume $\mu(G) \geq 2$. Then from the above proposition we infer that $a_G(1) \geq 1/2$. On the other hand by Theorem C

$$\frac{a_G(\lambda)}{a_G(\mu)\, a_G(\lambda-\mu)} \geq \binom{\lambda}{\mu}^s, \qquad 0 \leq \mu \leq \lambda,$$

where $s = \mu(G) - 1 \geq 1$. Our lemma applies to give:

Theorem D. *Under the assumption* $\mu(G) \geq 2$ *the function* b_G *is strictly increasing. Moreover for* $\lambda \geq 1$

$$b_G(\lambda+1) - b_G(\lambda) > m_G\, a_G(1)^\lambda\, (\lambda!)^s\, q(\lambda),$$

where $s = \mu(G) - 1$ *and* $q(\lambda) \in \mathbf{Q}[\lambda]$ *is the polynomial of degree* $\mu(G)$ *given by*

$$q(\lambda) = a_G(1)\, (\lambda+1)^{\mu(G)} - 2\, a_G(1)\, (\lambda+1) - \lambda.$$

Furthermore by Corollary C'

$$A_\lambda = \sum_{\mu=1}^{\lambda-1} \frac{a_G(\mu)\, a_G(\lambda-\mu)}{a_G(\lambda)} = O(\lambda^{-s}),$$

where $s = \mu(G) - 1 \geq 1$. In particular $A_\lambda \to 0$. Applying Wright's criterion (9) and evaluating the function $m_G\, \lambda\, a_G(\lambda)$ via (3) yields:

Theorem E. *For a finitely generated virtually free group* G *of rank* $\mu(G) \geq 2$ *the function* $b_G(\lambda)$ *is asymptotically equal to*

$$m_G\, \lambda\, a_G(\lambda) = m_G\, \lambda\, \frac{\prod_{e\in E} ((\lambda m_G/|G(e)|)!\; |G(e)|^{\lambda m_G/|G(e)|})}{\prod_{v\in V} ((\lambda m_G/|G(v)|)!\; |G(v)|^{\lambda m_G/|G(v)|})},$$

where the right-hand side is formed with respect to a decomposition of G *in terms of a finite graph* (G(-),Y) *of finite groups. More precisely we have*

$$1 = \frac{b_G(\lambda)}{m_G\, \lambda\, a_G(\lambda)} + O(\lambda^{-s}),$$

where $s = \mu(G) - 1$.

In the case $\mu(G) = 1$ i.e. G virtually infinite-cyclic the sequence $b_G(\lambda)$ is easily calculated, in fact $b_G(\lambda)$ is constant for these groups. This observation together with Theorem D finally establishes:

Corollary D'. *For a finitely generated virtually free group* G *the following conditions are equivalent:*

(i) G *contains a non-abelian free subgroup.*

(ii) *The function* b_G *is strictly increasing.*

(iii) *The function* b_G *is unbounded.*

References

[D] I M S Dey, Schreier systems in free products, *Proc. Glasgow Math. Soc.* **7** (1965), 61-79.

[H] M Hall, Subgroups of finite index in free groups, *Canad. J. Math.* **1** (1949), 187-190.

[I] S G Ivanov, Schreier systems in a free product of two groups with an amalgamated subgroup, *Math. Zap. Ural. Un-ta* **9** (1975), 13-33.

[KPS] A Karrass, A Pietrowski & D Solitar, Finite and infinite cyclic extensions of free groups, *J. Austral. Math. Soc.* **16** (1973), 458-466.

[M] Th Müller, *Kombinatorische Aspekte endlich erzeugter virtuell freier Gruppen* (Ph.D. Thesis, Universität Frankfurt am Main, 1989).

[N] M Newman, Asymptotic formulas related to free products of cyclic groups, *Math. Comp.* **30** (1976), 838-846.

[Se] J-P Serre, *Trees* (Springer-Verlag, 1980).

[St] J Stallings, On torsion-free groups with infinitely many ends, *Ann. of Math.* **88** (1968), 312-334.

[W] E M Wright, A relationship between two sequences, *Proc. London Math. Soc.* (3) **17** (1967), 296-304.

PROBLEMS IN LOOP THEORY FOR GROUP THEORISTS

MARKKU NIEMENMAA

University of Oulu, Finland

1. Introduction

The set of all left and right translations of a loop generates a group called the multiplication group of the loop. It is a well known fact that the concept of multiplication groups can very successfully be used in order to investigate loops. In this brief survey we shall proceed along two main lines: (1) the characterization of multiplication groups in purely group theoretic terms and (2) the structure and order of the multiplication group in the case that we have a finite centrally nilpotent loop. For the details of the results and their proofs we refer to [1] and [2].

2. Characterization of multiplication groups of loops

Let Q be a loop (thus Q is a groupoid with unique division and a neutral element). If $a \in Q$, then $L_a(x) = ax$ and $R_a(x) = xa$ define two permutations on Q called the left and the right translation. The set of all left and right translations generates a permutation group M(Q) and we say that M(Q) is the multiplication group of Q. Further, we denote by I(Q) the stabilizer of the neutral element. This permutation group is called the inner mapping group of Q. It is not difficult to see that I(Q) = 1 if and only if Q is an abelian group. If Q is a group, then I(Q) consists of the inner automorphisms of Q. Now put $A = \{L_a : a \in Q\}$ and $B = \{R_a : a \in Q\}$. Clearly, A and B are left (and right) transversals of I(Q) in M(Q).

Now we shall briefly introduce some notation. If G is a group and $H \leq G$, then H_G denotes the core of H in G. If A and B are two left transversals to H in G and $a^{-1}b^{-1}ab \in H$ whenever $a \in A$ and $b \in B$, then we say that A and B are H-connected transversals in G.

Our basic result is:

Theorem 1. *A group* G *is isomorphic to the multiplication group of a loop if and only if there exist a subgroup* H *satisfying* $H_G = 1$ *and* H-*connected transversals* A *and* B *satisfying* $G = \langle A,B \rangle$.

In the preceding theorem H is naturally in the role of I(Q) and A and B correspond to the transversals $\{L_a : a \in Q\}$ and $\{R_a : a \in Q\}$. On the other hand, we can define a binary operation in the set K of all left cosets of H in such a way that K is a loop and M(K) is isomorphic to G.

Our next result is purely group theoretical and it deals with H-connected transversals.

Theorem 2. *Let* H *be a cyclic subgroup of a finite group* G. *Then* $G' \le H$ *if and only if there exists a pair* A, B *of* H*-connected left transversals in* G *such that* $G = \langle A,B \rangle$.

Since the core of I(Q) in M(Q) is trivial, the combination of Theorems 1 and 2 yields:

Theorem 3. *If* Q *is a finite loop such that* I(Q) *is a cyclic group, then* Q *is an abelian group.*

We have been able to prove that Theorem 3 also holds in the case that Q is infinite and I(Q) is a cyclic p-group. However, the general case is open and we can state our first problem.

Problem. Assume that Q is a loop and I(Q) a cyclic group. Does it follow that Q is an abelian group?

By using the properties of I(Q) we can give examples of finite groups which are not isomorphic to multiplication groups of loops. Here is one example: if G is a finite nonabelian group with the property that every noncyclic subgroup is normal, then G is not isomorphic to the multiplication group of a loop (a complete characterization of such groups can be found in [3]).

We end this section with a very general problem.

Problem. Which groups are isomorphic to multiplication groups of loops?

3. Centrally nilpotent loops

In this section we assume that Q is a finite loop. The centre Z(Q) consists of all elements a which satisfy the equations ax.y = a.xy, xa.y = x.ay, xy.a = x.ya and xa = ax whenever x and y are elements of Q. Clearly, Z(Q) is an abelian subgroup of Q and Z(Q) and Z(M(Q)) are isomorphic.

Now assume that the loop Q/Z(Q) has order r. If we define

$$K = K(Z(Q)) = \{P \in I(Q) : P(x) \in xZ(Q) \quad \text{for all} \quad x \in Q\},$$

then K is a normal subgroup of I(Q) and I(Q/Z(Q)) is isomorphic to I(Q)/K. Furthermore, K is isomorphic to a subgroup of Z(Q) × ... × Z(Q) where Z(Q) appears r-1 times.

We now define a series of normal subloops of Q as follows: $Z_0 = 1$, $Z_1 = Z(Q)$ and $Z_i/Z_{i-1} = Z(Q/Z_{i-1})$. If there is an integer c such that Z_{c-1} is a proper subloop of Q but $Z_c = Q$ then we say that Q is centrally nilpotent of class c. If we put $K_0 = 1$, $K_1 = K$ and K_i/K_{i-1} isomorphic to $K(Z_i/Z_{i-1})$ then we have an ascending series of normal subgroups K_i of I(Q) and $K_{c-1} = I(Q)$ if Q is centrally nilpotent of class c. The normal subgroups K_i have the following two properties: $I(Q)/K_i$ is isomorphic to $I(Q/Z_i)$ and K_i/K_{i-1} is isomorphic to a subgroup of

$$(Z_i/Z_{i-1}) \times \ldots \times (Z_i/Z_{i-1})$$

where Z_i/Z_{i-1} appears r_i -1 times, r_i = the order of Q/Z_i.

Since the groups K_i/K_{i-1} are all abelian, we can state our first result in this section.

Theorem 4. *If* Q *is a finite centrally nilpotent loop, then* I(Q) *is solvable. If* Q *has order* n, *then the orders of* I(Q) *and* M(Q) *divide some power of* n.

From Theorem 4 it follows that if Q is a finite centrally nilpotent p-loop (the order of Q is a power of a prime p), then I(Q) and M(Q) are finite p-groups.

We shall now demonstrate how to use our results in order to determine the structure of I(Q) for a loop Q which has order 6. The loop table is as follows:

1	2	3	4	5	6
2	1	4	3	6	5
3	4	5	6	2	1
4	3	6	5	1	2
5	6	1	2	3	4
6	5	2	1	4	3

Now Z(Q) = {1,2} and K = I(Q). We also know that I(Q) is isomorphic to a subgroup of Z(Q) × Z(Q). By Theorem 3, I(Q) is not a cyclic group and it follows that I(Q) is isomorphic to the Klein four group. Clearly, M(Q) has order 24.

We finally set ourselves to the task of showing the relation between the central nilpotency of Q and the structure of M(Q). If we put $I_0 = I(Q)$ and put $I_i = N_{M(Q)}(I_{i-1})$, then

$$I_i = \{R_x U : x \in Z_i \quad \text{and} \quad U \in I(Q)\}.$$

It follows that Q is centrally nilpotent if and only if there exists a natural number d such that $I_d = M(Q)$. Now I_{i-1} is normal in I_i and the factor group I_i/I_{i-1} is abelian. If Q is centrally nilpotent then, by Theorem 4, M(Q) is solvable. On the other hand, if M(Q) is a finite nilpotent group then Q is centrally nilpotent. In the case of p-loops we have a very interesting result.

Theorem 5. *Let* Q *be a* p-*loop. Then* Q *is centrally nilpotent if and only if* M(Q) *is a* p-*group.*

References

1. R H Bruck, Contributions to theory of loops, *Trans. Amer. Math. Soc.* **60** (1946), 245-354.

2. T Kepka & M Niemenmaa, On multiplication groups of loops, *J. Algebra*, to appear.

3. A D Ustozhanikov, Konechnye gruppy s invariantymi netsiklicheskimi podgruppami, *Mat. Zapicki, Tetradj* **1** (1967).

SYLOW THEORY OF CC-GROUPS: A SURVEY

J OTAL & J M PEÑA

Universidad de Zaragoza, 50009 Zaragoza, Spain

1. Introduction

Groups with Černikov conjugacy classes or CC-*groups* were considered by Polovickiĭ [11] as an extension of the concept of an FC-group, that is, a group in which every element has only a finite number of conjugates. A group G is said to be a CC-*group* if $G/C_G(x^G)$ is a Černikov group for each $x \in G$ (see also [12, 4.36]).

In [1] and [7] a classical Sylow theory for CC-groups was initiated. The local conjugacy of the Sylow p-subgroups of a CC-group G was established as was the characterization of when they are conjugate in terms of their number and of internal properties of the group G. We now present a summary of the generalization of the above results to an arbitrary set π of primes. We also consider the usual ingredients of a more complete Sylow theory such as Sylow bases or Carter subgroups, thought of as an extension of the FC-case: see [4], [15], [16], [17] and [18]. The details of the contents of this note can be found in [9] and this, being a survey article, does not contain proofs, except as an illustration.

Throughout our notation is standard and is taken from [12] and [18], to which we refer for the basic definitions and the setting of the problems we are considering. Some parts of the proofs of our results, as we present them here, depend on the interpretation of a residually Černikov group as a co-Černikov group, following the topological approach set out and developed by Dixon [2]. This allows us to make use of a version of the inverse limit theorem (see [2, 1.1]), which is useful in establishing the local conjugacy of some elements of this theory, as well as showing that locally inner automorphisms of subgroups and quotients of CC-groups are induced by locally inner automorphisms of the whole group ([9, 2.2]). On the other hand, other proofs, especially those concerning Carter subgroups and related ideas, are made by reducing the problem to FC-groups and even to U-groups. The idea is to find adequate FC-quotients of a CC-group or to reduce the problems to locally finite groups with conjugate Sylow subgroups (see (3.4) and (4.2) below).

2. Sylow subgroups

The existence of Sylow π-subgroups of a group G is trivial since they are exactly the maximal π-subgroups of G. To study their local conjugacy we say that a CC-group G is a C_π-*group* if, in each Černikov subgroup H of G, the Sylow π-subgroups of H are conjugate in H. Clearly any CC-group G is a C_p-group, for every prime p, and a locally soluble CC-group G is a C_π-group, for every π. Thus we have:

Theorem 2.1. *Let π be a set of primes and let* G *be a* C_π-*group.*

(a) *The Sylow π-subgroups of* G *are locally conjugate.*

(b) *The following conditions are equivalent:*

(1) *The Sylow π-subgroups of* G *are conjugate in* G.

(2) G *has a countable number of Sylow π-subgroups.*

(3) $G/O_\pi(G)$ *satisfies the minimal condition for π-subgroups.*

(4) *The Sylow π-subgroups of* $G/O_\pi(G)$ *are Černikov groups.*

(5) *The Sylow π-subgroups of* $G/O_\pi(G)$ *are finite groups.*

(6) *The torsion subgroup of* G *is a finite extension of a π'-extension of* $O_\pi(G)$.

(7) G *is a finite extension of a π^*-extension of* $O_\pi(G)$.

The proof of the local conjugacy is an application of the technique of the inverse limit theorem and is the same as that in the case where π consists of a unique prime ([1, Theorem 1]). Concerning the characterization of conjugation, we first note that condition (5) is identical to that of the FC-case and gives a strong finiteness condition on the structure of a C_π-group. In general, if G is a CC-group and $P \in Syl_\pi(G)$, then $P/O_\pi(G)$ is an FC-group ([9, 3.3]). Moreover, the conditions (6) and (7) are similar to those of [13] and the present form of the latter is due to Mike J Tomkinson. The equivalence between (3), (4) and (5) follows because $O_\pi(G)$ contains any radicable π-subgroup of G ([9, 3.3]) and the equivalence between (5), (6) and (7) can be shown as in [7]. Clearly (5) $\Rightarrow$ (1) and (5) $\Rightarrow$ (2). To finish the proof we may show (1) $\Rightarrow$ (2) as in [1, Theorem 2], prove (1) $\Rightarrow$ (5) and (2) $\Rightarrow$ (5) as in [7], or use one of the latter and prove directly (2) $\Rightarrow$ (3). With the exception of the proof of (2) $\Rightarrow$ (5), which is based on an idea due to B Hartley, the other proofs use deeply the topological approach mentioned in the introduction. We finally remark that the proof of this result as given in [9] is slightly different.

3. Sylow bases, complement systems, basis normalizers and Carter subgroups

As usual, *a Sylow basis of a group* G is a set $S = \{S_p\}$ of Sylow p-subgroups of G, one for each prime p, such that the subgroup $< S_p \mid p \in \pi >$ is a π-group, for each set π of primes. In order to state the existence and the local conjugacy of a Sylow basis, we follow the approach originally due to P Hall. To do that we say that *a Sylow complement system of a group* G is a set $K = \{S_{p'}\}$ of Sylow p'-subgroups of G, one for each prime p. Thus, it is clear that G has a complement system. Moreover, a Sylow basis $S = \{S_p\}$ of a CC-group G determines a Sylow complement system $K = \{S_{p'}\}$ of G given by

$$S_{p'} = < S_q \mid q \in p' >$$

for each p. As in the finite case we have a bijection between Sylow bases and complement systems, which allows us to extend results of Gol'berg [4] and Stonehewer [15] to CC-groups (see [18, 5.22]).

Lemma 3.1. *Let* $K = \{S_{p'}\}$ *be a complement system of a locally soluble* CC-*group* G. *If we define*

$$S_p = \cap\{S_{q'} \mid q \neq p\},$$

then S_p *is a Sylow* p-*subgroup of* G *and* $S = \{S_p\}$ *is a Sylow basis of* G. *Moreover, the above correspondence between Sylow bases and complement systems is one to one.*

Theorem 3.2. *A* CC-*group* G *has a Sylow basis if and only if* G *is locally soluble and, in this case, any two Sylow bases of* G *are locally conjugate in* G.

It follows from (3.2) that the complement systems of G are locally conjugate. Another consequence of the relationship between Sylow bases and complement systems is that we have

$$\cap\{N_G(S_p) \mid p \text{ prime}\} = \cap\{N_G(S_{p'}) \mid p \text{ prime}\},$$

whenever $S = \{S_p\}$ is a Sylow basis and $K = \{S_{p'}\}$ is a complement system of a locally soluble CC-group G which corresponds in the above bijection. This subgroup, D say, is called *the basis normalizer associated with* S. Clearly these normalizers are also locally conjugate.

By definition, a *Carter subgroup of a CC-group* G is a self-normalizing locally nilpotent subgroup of G. In [9, 4.6], we showed that a Carter subgroup of a periodic locally soluble CC-group G is exactly a locally nilpotent projector of G. This is fundamental in what follows. The key point in establishing the existence and the local conjugacy of Carter subgroups is to study the poly-locally nilpotent case, which is also a key to dealing with the characterization of all the elements we are describing.

Theorem 3.3. *Let* G *be a periodic locally nilpotent-by-locally nilpotent* CC-*group. Then we have :*

(1) *The Carter subgroups of* G *are exactly the basis normalizers of* G.

(2) *There is a bijection between the Sylow bases of* G *and the basis normalizers of* G *which is stable under conjugation.*

Theorem 3.4. *A periodic locally soluble* CC-*group* G *has Carter subgroups and any two Carter subgroups of* G *are locally conjugate in* G.

It is worthwhile giving the proof of (3.4) to understand how one can reduce the problem to FC-groups. We think of Carter subgroups as locally nilpotent projectors so that we shall make use of the properties of projectors in the next argument. Let R be the radicable part of G. By [7, 2.1], R is abelian and G/R is an FC-group. By [18, 6.19], we may choose a locally nilpotent projector L/R of G/R. Then L is abelian-by-locally nilpotent so that (3.2) and (3.3) assure that L has a locally nilpotent projector, C say, and it is clear that C is then a Carter subgroup of G.

Now let C_1 and C_2 be two Carter subgroups of G. Then C_1R/R and C_2R/R are locally nilpotent projectors of G/R so that, by [18, 6.19], there is some $\varphi \in \mathrm{Linn}(G/R)$ such that

$$(C_1R/R)^{\varphi} = C_2R/R,$$

and we may extend φ to an element $\theta \in \mathrm{Linn}(G)$. Thus $(C_1)^{\theta}$ and C_2 are Carter subgroups of C_2R and, since C_2R is abelian-by-locally nilpotent, by (3.2) and (3.3), $(C_1)^{\theta}$ and C_2 are locally conjugate in C_2R and hence in G.

4. The theorem of characterization of conjugacy

This section is devoted to stating the "conjugacy theory" inherent in the theory of Sylow which we have described. The situation is very similar to the corresponding theory of periodic FC-groups ([16], [18]).

Some parts of the proof are shown by investigating a CC-group modulo its radicable part or its Hirsch-Plotkin radical. We shall obtain stuctural properties of the corresponding quotients and make use of the characterization of U-groups done by B Hartley in [5] and [6] as well as some properties of these groups given in [3]. To translate this information back to G we shall need an auxiliary result which we state before the main theorem.

Lemma 4.1. *For a periodic* CC-*group* G *the following are equivalent:*

(1) G *is locally nilpotent-by-finite.*

(2) G *is locally nilpotent-by-Černikov.*

(3) G *is Černikov-by-locally nilpotent.*

Theorem 4.2. *For a periodic locally soluble* CC-*group* G *the following conditions are equivalent:*

(1) *The Sylow bases of* G *are conjugate.*

(2) G *has a countable number of Sylow bases.*

(3) *The Sylow complement systems of* G *are conjugate.*

(4) G *has a countable number of Sylow complement systems.*

(5) *For each set of primes* π, *the Sylow* π*-subgroups of* G *are conjugate.*

(6) *For each set of primes* π, G *has a countable number of Sylow* π*-subgroups.*

(7) *The Carter subgroups of* G *are conjugate.*

(8) G *has a countable number of Carter subgroups.*

(9) *The basis normalizers of* G *are conjugate.*

(10) G *has a countable number of basis normalizers.*

(11) G *is locally nilpotent-by-finite.*

We shall give a sketch of the proof, whose details can be found in full in [9, 4.12]. In what follows we denote by H the Hirsch-Plotkin radical of G and by R the radicable part of G. Thus R is abelian, R ≤ H, G/R and G/H are FC-groups and G/H is residually finite ([7, 2.1] and [8, Theorem B]).

The equivalences (1) ⇔ (3) and (2) ⇔ (4) are a clear consequence of (3.1) and the equivalence (5) ⇔ (6) of (2.1). (1) ⇒ (9), (2) ⇒ (10) and (11) ⇒ (2) are immediate and (3) ⇒ (5) and (4) ⇒ (6) follow from the fact that a Sylow π-

subgroup of G can always be obtained as the intersection of some members of a certain complement system of G.

The proof of (5) or (6) ⇒ (1) and (7) is an example of how to reduce the question to U-groups. It is clear that each subgroup of G also satisfies (6) (and so (5)). Then we may apply [6, Theorem E] to deduce that G is a U-group. Therefore (1) and (7) follow from [3, 2.10 and 5.4].

The proof of (1) ⇒ (11) is the heart of (4.2) (see [9, 4.12] for another shorter proof). We know that (1) and (6) are equivalent so that every subgroup and every factor of G satisfy (1) and (6). In particular G is again a U-group. Since G/H is an FC-group, by [16, Theorem D], we have that G/H is locally nilpotent-by-finite. Thus, in order to show that G/H is finite, it is shown that we may assume that G is countable, G/H is locally nilpotent and Z(G) = 1. By [10, Theorem 6] G can be viewed as a subgroup of the direct product D of a countable family $\{C_n \mid n \geq 1\}$ of Černikov groups. Fix a Sylow basis $S = \{S_p\}$ of G. If J is a finite set of natural numbers and $K_J = \mathrm{Dr}\ \{C_n \mid n \in J\}$, then

$$S \cap K_J = \{S_p \cap K_J\}$$

is a Sylow basis of $G \cap K_J$. The hardest part of this proof is to show that there exists an integer m such that, whenever J is a finite set and $J \cap \{1,2,\ldots,m\} = \emptyset$, then $S_p \cap K_J$ is normal in G for every p. If we define

$$T = \mathrm{Dr}\ \{C_n \mid n > m\},$$

then $S_p \cap T$ is normal in G so $G \cap T$ is locally nilpotent and hence $G \cap T \leq H$. Clearly $G/G \cap T$ is Černikov and so is G/H. By (4.1) G/H is therefore finite.

It is interesting to say something about the proof of the statements (7), (8), (9) or (10) ⇒ (11) because it is deeply used in the reduction to FC-groups. First of all, G/R satisfies (i) provided G satisfies (i), where i = 7, 8, 9 or 10. Here G/R is an FC-group so that, by [18, 4.21 and 4.25], in the cases (8) or (10), we may assume that G/R has a finite class of conjugacy of Carter subgroups or basis normalizers. In any case we are in a position to apply [16, Theorem D] and [18, 6.31] to deduce that G/R is finite-by-locally nilpotent. If F is a normal subgroup of G containing R such that F/R is finite and G/F is locally nilpotent, it can be shown that F' is Černikov and G/F' is abelian-by-locally nilpotent so that, by (3.3), we have that G/F' satisfies (1) or (2). We have already shown that G/F' is then locally nilpotent-by-finite. Then G has a normal subgroup L of finite index such that L is Černikov-by-locally nilpotent. By (4.1) L is locally nilpotent-by-finite and therefore so is G.

Assume (2). Since (2) and (6) are equivalent G is a U-group. We note that G is also locally nilpotent-by-finite. Clearly G has a countable number of basis normalizers. By [14, 3.6], every Carter subgroup of G contains at least one basis normalizer and every basis normalizer is contained in at least one Carter subgroup of G, which is unique by [3, 5.10]. Therefore the number of Carter subgroups of G is countable.

Acknowledgement

This research was supported by DGICYT (Spain) PS88-0085.

References

1. J Alcázar & J Otal, Sylow subgroups of groups with Černikov conjugacy classes, *J. Algebra* **110** (1987), 507-513.
2. M R Dixon, Some topological properties of residually Černikov groups, *Glasgow Math. J.* **23** (1982), 65-82.
3. A D Gardiner, B Hartley & M J Tomkinson, Saturated formations and Sylow structure in locally finite groups, *J. Algebra* **17** (1971), 177-211.
4. P A Gol'berg, Sylow p-subgroups of locally normal groups, *Mat. Sb.* **19** (1946), 451-460.
5. B Hartley, Sylow subgroups of locally finite groups, *Proc. London Math. Soc.* (3) **23** (1971), 159-192.
6. B Hartley, Sylow theory in locally finite groups, *Comp. Math.* **23** (1972), 263-280.
7. J Otal & J M Peña, Characterizations of the conjugacy of Sylow p-subgroups of CC-groups, *Proc. Amer. Math. Soc.* **106** (1989), 605-610.
8. J Otal & J M Peña, Locally nilpotent injectors of CC-groups, in *Contribuciones Matemáticas em Homenaje al Profesor D Antonio Plans* (Universidad de Zaragoza, 1990), 233-238.
9. J Otal & J M Peña, Sylow theory of CC-groups, *Rend. Sem. Mat. Univ. Padova*, to appear.
10. Ya D Polovickiĭ, On locally extremal groups and groups with the condition of π-minimality, *Soviet Mat. Dokl.* **2** (1961), 780-782.
11. Ya D Polovickiĭ, Groups with extremal classes of conjugate elements, *Sibirsk. Mat. Ž.* **5** (1964), 891-895.
12. D J S Robinson, *Finiteness conditions and generalized soluble groups* (Springer-Verlag, 1972).
13. E Schenkman, Groups with a finite number of Sylow Λ-subgroups, *Proc. Amer. Math. Soc.* **57** (1976), 205.
14. S E Stonehewer, Abnormal subgroups of a class of periodic locally soluble groups, *Proc. London Math. Soc.* (3) **14** (1964), 520-536.

15. S E Stonehewer, Locally soluble FC-groups, *Arch. Math.* **26** (1965), 158-177.

16. S E Stonehewer, Some finiteness conditions in locally soluble groups, *J. London Math. Soc.* (3) **19** (1969), 675-708.

17. M J Tomkinson, Formations of locally soluble FC-groups, *Proc. London Math. Soc.* (3) **19** (1969), 675-708.

18. M J Tomkinson, *FC-groups* (Pitman, 1984).

ON THE MINIMAL NUMBER OF GENERATORS OF CERTAIN GROUPS

LUIS RIBES

Carleton University, Ottawa, Ont K1S 5B6, Canada

KIEH WONG

Indiana University, South Bend, Indiana 46634, USA

1. Introduction

Let H_1 and H_2 be finite groups. The results in this paper are intended as a contribution to the investigation of the following problem: does there exist a finite group G containing H_1 and H_2, generated by them, and such that

$$d(G) = d(H_1) + d(H_2),$$

where d(G) denotes the minimal number of generators of the group G? This problem, in turn, arises from the theory of profinite groups when trying to prove, for free profinite products of profinite groups, an analogue of the Grushko-Neumann theorem for free products of abstract groups ([2], [6]; see Section 2). If, for example, H_1 and H_2 are finite p-groups, for a fixed prime p, then obviously G can be taken to be their direct product $H_1 \times H_2$; and consequently the analogue of the Grushko-Neumann theorem holds for free pro-p-products of pro-p-groups (cf. [5, Prop. 2.9]).

We restrict our attention to constructions involving direct products and wreath products. On the other hand, we allow the groups to be infinite in some cases, as long as the constructions used yield finite groups whenever we start with finite groups (note that unless we impose such a condition, one could define the group G above, in the infinite case, to be the free product of H_1 and H_2). Also, we give bounds for the minimal number of generators of certain groups that appear naturally in our constructions. Throughout the paper our aim is to deal with explicit constructions and to describe explicit generators and bounds, whenever possible.

In Section 2 we give a series of equivalent conditions, in the context of profinite groups, to the problem mentioned above. In Section 3, we describe bounds for the minimal number of generators of direct products and wreath products of groups whose factors admit minimal sets of generators with mutually coprime orders. In Section 4, we compute the minimum number of generators of groups of the form $G \wr C_n$, $C_n \wr D_m$, $D_n \wr D_m$, $C_n \wr Q_m$, and $Q_n \wr Q_m$ where C_n is cyclic of order n, G is for example finite nilpotent, D_m denotes the dihedral group of order 2m, and Q_m is the dicyclic group of order 2m (in this last case m is even).

2. Motivation

Let $\mathbf{C}$ be a class of finite groups closed under the operations of taking subgroups, quotients and extensions, and let G be a pro-$\mathbf{C}$-group, i.e., G is a projective limit of groups in $\mathbf{C}$. The group G carries with it a natural topology, which is compact, Hausdorff and totally disconnected (see [8] or [7] for details). One says that a subset X of G generates G (as a pro-$\mathbf{C}$-group) if the abstract subgroup of G generated by X is dense in G. We denote by d(G) the smallest cardinality of a generating set of G. If X is a generating set of G, and $|X| = d(X)$, we refer to X as a minimal generating set of G.

Lemma 2.1. *Let* $\{G_i \mid \phi_{ij}\}$ *be a projective system of pro-*$\mathbf{C}$*-groups over a directed set* $(1,\leq)$, *with each* ϕ_{ij} *an epimorphism and consider the pro-*$\mathbf{C}$*-group* $G = \varprojlim G_i$. *Then* $d(G) = \max d(G_i)$.

Proof. Denote by $\phi_i : G \to G_i$ the canonical projection for each $i \in 1$. If X is a generating set of G, then $\phi_i(X)$ generates G_i, and so $d(G) \geq \max d(G_i)$. In particular, if $\max d(G_i)$ is infinite, then so is d(G). Suppose then that $\max d(G_i) = n < \infty$. Let C_i denote the subspace of the n-th cartesian power G_i^n of the group G_i consisting of those n-tuples that generate G_i. By our assumption $C_i \neq \emptyset$, and on the other hand C_i is compact (to see this, express G_i as a projective limit of finite groups G_{ij}, and note that C_i can then be expressed as an inverse limit of finite subsets of G_{ij}^n). Now if $i \geq j$, then $\phi_{ij}(C_i) \subseteq C_j$, i.e., $\{C_i \mid \phi_{ij}\}$ is a projective system of compact, non-empty sets. Therefore,

$$C := \varprojlim C_i \neq \emptyset$$

([1, Prop. 8, p.102]). Note that C is a subset of G^n. Let $(x_1,\ldots,x_n) \in C$. Then $\{x_1,\ldots,x_n\}$ generates G, since $\phi_i\{x_1,\ldots,x_n\}$ generates G_i for each $i \in I$. Thus, $d(G) \leq \max d(G_i)$.

Let H_1 and H_2 be pro-$\mathbf{C}$-groups. Consider their free pro-$\mathbf{C}$-product $G = H_1 \amalg H_2$ (i.e., their coproduct in the category of pro-$\mathbf{C}$-groups). An open question in the theory of profinite groups is whether or not $d(G) = d(H_1) + d(H_2)$, as an analogue to the Grushko-Neumann theorem for abstract free products of groups. The following results indicate how this question relates to questions in combinatorial group theory and the theory of finite groups. Part (iv) is the main motivation for our results in the paper.

Proposition 2.2. *The following assertions are equivalent:*

(i) *For every pair* H_1 *and* H_2 *of pro-*$\mathbf{C}$*-groups,*

$$d(H_1 \amalg H_2) = d(H_1) + d(H_2);$$

(ii) *For every pair* H_1 *and* H_2 *of finite groups in* $\mathbf{C}$,

$$d(H_1 \amalg H_2) = d(H_1) + d(H_2);$$

(iii) *For every pair of* H_1 *and* H_2 *of finite groups in* $\mathbf{C}$*, there is a finite quotient* H *in* $\mathbf{C}$ *of the abstract free product* $H_1 * H_2$*, such that* $d(H) = d(H_1) + d(H_2)$;

(iv) *For every pair* H_1 *and* H_2 *of finite groups in* $\mathbf{C}$*, there is a finite group* H *in* $\mathbf{C}$ *such that* H_1 *and* H_2 *are subgroups of* H*, the group* H *is generated by* H_1 *and* H_2*, and* $d(H) = d(H_1) + d(H_2)$.

Proof. Observe that (ii) is a particular case of (i). To prove that (ii) implies (i), consider finitely generated pro-$\mathbf{C}$-groups H_1 and H_2, and note that

$$H_1 \amalg H_2 = \varprojlim [H_1/(U \cap H_1) \amalg H_2/(U \cap H_2)],$$

where U runs through open normal subgroups of G (cf., e.g., [4, Lemman 4.2]). Using Lemma 2.1, we may assume (restricting to a cofinal set of U's if necessary) that for all U one has that

$$d(H_i) = d(H_i/(U \cap H_i)), \; i = 1,2.$$

By Lemma 2.1 again,

$$d(H_1 \amalg H_2) = d[H_1/(\bar{U} \cap H_1) \amalg H_2/(\bar{U} \cap H_2)],$$

for some open normal subgroup $\bar{U}$ of G. Hence by (ii),

$$d(H_1 \amalg H_2) = d(H_1/(\bar{U} \cap H_1)) + d(H_2/(\bar{U} \cap H_2)) = d(H_1) + d(H_2).$$

To show that (iii) follows from (ii), note that $H_1 \amalg H_2 = \varprojlim (H_1 * H_2/N)$ where N runs through the normal subgroups of $H_1 * H_2$ of finite index. By Lemma 2.1, there exists some open normal subgroup $\bar{N}$ of $H_1 * H_2$ such that

$$\bar{N} \cap H_i = 1 \; (i = 1,2)$$

and $d(H_1 \amalg H_2) = d(H_1 * H_2/\bar{N})$, and therefore by (ii),

$$d(H_1 * H_2/\bar{N}) = d(H_1) + d(H_2).$$

Clearly (iii) implies (iv). Finally, assume (iv); then there is an obvious epimorphism of $H_1 \amalg H_2$ onto H, so that

$$d(H_1 \amalg H_2) \geq d(H) = d(H_1) + d(H_2);$$

hence $d(H_1 \amalg H_2) = d(H_1) + d(H_2)$.

The next result gives some evidence towards an eventual proof that in fact the equivalent conditions of Proposition 2.2 always hold. It suggests that if one starts with big enough finite groups H_1 and H_2, then there is a group G as in condition (iv). We are grateful to W Herfort for his suggestions about this result.

Proposition 2.3. *Let* H_1 *and* H_2 *be finite groups in* $\mathcal{C}$, *and let* F_1 *and* F_2 *be free pro-*$\mathcal{C}$*-groups of rank* $d(H_1)$ *and* $d(H_2)$, *respectively. Then there exist open normal subgroups* $\bar{U}_1$ *and* $\bar{U}_2$ *of* F_1 *and* F_2 *respectively, such that for every open normal subgroup* $U_1 \leq \bar{U}_1$ *and* $U_2 \leq \bar{U}_2$ *of* F_1 *and* F_2 *respectively, one has that*

$$d(F_1/U_1) = d(H_1) \text{ and } d(F_2/U_2) = d(H_2),$$

and

$$d(F_1/U_1 \amalg F_2/U_2) = d(F_1/U_1) + d(F_2/U_2).$$

Proof. Observe first that $F_1 \amalg F_2$ is the free pro-$\mathcal{C}$-group of rank

$$d(H_1) + d(H_2).$$

Since

$$F_1 \amalg F_2 = \varprojlim [F_1/(V \cap F_1) \amalg F_2/(V \cap F_2)],$$

where V runs through the open normal subgroups of $F_1 \amalg F_2$ (see, e.g., [4, Lemma 4.2]), there exists some $\bar{V}$, such that

$$d(H_1) + d(H_2) = d(F_1 \amalg F_2) = d(F_1/(\bar{V} \cap F_1) \amalg F_2/(\bar{V} \cap F_2)),$$

by Lemma 2.1. Then let $\bar{U}_1 = \bar{V} \cap F_1$ and $\bar{U}_2 = \bar{V} \cap F_2$.

Remark 2.4. A more explicit way of proving a slightly less general variant of Proposition 2.3, is as follows. Let $x_1, \ldots, x_t$ be a minimal set of generators of H_1.

Choose a prime number p which does not divide the order of any of the x_i's, and such that $C_p \in \mathbf{C}$. Then clearly, $d(H_1 \times C_p^t) = d(H_1)$. If $d(H_2) = s$, choose the prime p so that also $d(H_2 \times C_p^s) = d(H_2)$. Then plainly,

$$d((H_1 \times C_p^t) \times (H_2 \times C_p^s)) = d(H_1 \times C_p^t) + d(H_2 \times C_p^s).$$

3. Preliminaries and first results

The main constructions we are dealing with in this paper are the direct and the wreath products of groups. We begin by stating some elementary properties involving these operations, that will be used freely throughout this paper.

Lemma 3.1. (i) *Let* G *and* H *be groups, and* K *a normal subgroup of* G. *Then there exists an epimorphism* $G \wr H \to (G/K) \wr H$, *defined by*

$$(g_1,\ldots,g_n)h \mapsto (g_1K,\ldots,g_nK)h,$$

where $g_i \in G$, $h \in H$.

(ii) *Let* G *be any group and* H *a finite group. Then there exists an epimorphism* $G \wr H \to G/G' \times H/H'$, *defined by*

$$(g_1,\ldots,g_n)h \mapsto (g_1 \ldots g_nG')(hH'),$$

where $g_i \in G$, $h \in H$.

Lemma 3.2. *Let* A *be an abelian group,* G *any group and* H *a finite normal subgroup of* G. *Then there is an epimorphism* $\phi : A \wr G \to A \wr (G/H)$.

Proof. Consider an element $\alpha g \in A \wr G$, where $\alpha : G \to A$ is a function, and $g \in G$. Define $\phi(\alpha g) = \bar{\alpha}\bar{g}$, where $\bar{\alpha} : G/H \to A$ is given by

$$\bar{\alpha}(gH) = \prod_{h \in H} \alpha(xh),$$

and $\bar{g} = gH$.

Let A be a finitely generated abelian group, and consider its canonical decomposition $A = A_1 \oplus \ldots \oplus A_t \oplus \mathbf{Z}^r$, where each A_i is finite cyclic, $t \geq 0$,and for each i the order of A_i divides the order of A_{i+1}. Note that $d(A) = t+r$. If B is another finitely generated abelian group with canonical decomposition

$$B = B_1 \oplus \ldots \oplus B_s \oplus \mathbf{Z}^u,$$

we say that A and B are *not coprime* if either $t, s \geq 1$ and A_1 and B_1 have not coprime orders, or if $t+s = 0$.

If G is a group, we say that it admits a *coprime set of generators*, if there exists a set $x_1,...,x_t$ of generators of G, each of finite order, and such that their orders are mutually coprime. Note that the abelianized group G/G' of such a group is cyclic of finite order.

Proposition 3.3. *Let* G *and* H *be finitely generated groups such that* $d(G) = d(G/G')$ *and* $d(H) = d(H/H')$.

(i) *If* G *and* H *are abelian,* $d(G \times H) = d(G) + d(H)$ *if and only if* G *and* H *are not coprime;*

(ii) *if* G/G' *and* H/H' *are not coprime one always has* $d(G \times H) = d(G) + d(H)$; *and*

(iii) *if in addition* H *is finite, and if* G/G' *and* H/H' *are not coprime, then*

$$d(G \wr H) = d(G) + d(H).$$

Proof. (i) Suppose first that G and H are not coprime; then either one of them is torsion-free or there is a prime number p which divides the order of each of the finite factors in the canonical decomposition of G and of H; then there is an epimorphism from $G \times H$ onto the cartesian product of $d(G) + d(H)$ copies of Z/pZ. Therefore, $d(G \times H) = d(G) + d(H)$. Conversely, if G and H are coprime, then G and H contain some torsion and the first factors G_1 and H_1 of their canonical decompositions have relatively prime orders (see the notation in Section 3). But then $G_1 \times H_1$ is cyclic, and therefore

$$d(G \times H) \leq d(G) + d(H) - 1.$$

(ii) and (iii) follow from (i) and Lemma 3.1.

Proposition 3.4. *Let* G *be a finitely generated group with a coprime minimal set of generators, and let* H *be any group. If* $d(G) \leq |H|$, *then*

$$d(G \wr H) \leq d(H) + 1.$$

And in general, if H *is finite,* $d(G \wr H) \leq d(H) + \alpha$, *where* α *is the smallest natural number greater or equal to* $d(G)/|H|$.

Proof. Let $x_1,...,x_n$ be a set of coprime generators of G. Assume first that $d(G) \leq |H|$. Express $G \wr H = K \rtimes H$ as a semidirect product, where

$$K = \prod_{h \in H} G^h$$

is the direct product of $|H|$ copies of G. Choose n distinct elements $h(1),\dots,h(n)$ of H, and consider the n elements $x_1^{h(1)},\dots,x_n^{h(n)}$ of K. It is plain that these elements together with H generate G. On the other hand $< x_1^{h(1)},\dots,x_n^{h(n)} >$ is a cyclic group generated by the product $x_1^{h(1)} \dots x_n^{h(n)}$, since the elements $x_1^{h(1)},\dots,x_n^{h(n)}$ commute and have relatively prime orders. Thus $G \wr H \le d(H) + 1$. The second part is proved similarly.

Corollary 3.5. *Let* G *be a finitely generated imperfect group (i.e.,* $G' \neq G$) *with a coprime minimal set of generators* $x_1,\dots,x_n$, *and let* $m \ge 2$ *and* t *be natural numbers. Assume that* $tm \ge n$, *and that* m *is not relatively prime with the order of some* x_i. *Then* $d(G \wr C_m^t) = t+1$.

Proof. By Proposition 3.3, $d(G \wr C_m^t) \le t+1$. On the other hand, by Lemma 3.1, there exist epimorphisms

$$G \wr C_m^t \to G/G' \wr C_m^t \to G/G' \times C_m^t.$$

Finally, by assumption m is not coprime to the order of the cyclic group G/G', and thus $d(G/G' \times C_m^t) = t+1$, by Proposition 3.3.

Let r be a natural number and let G^r denote the cartesian product of r copies of a group G. In [9, Th. 2.2], it is shown that $d(G^r) = r\,d(G/G')$ provided G is a finitely generated imperfect group and $r \ge 2(2 + d(G)/d(G/G'))^2$. Our next proposition gives a sharper result, in the special case that G has a minimal set of generators of finite mutually coprime orders.

Proposition 3.6. *Let* G *be a finitely generated group that admits a coprime minimal set of generators. Then* $d(G^r) = d(G)$, *if* $r \le d(G)$; *and* $d(G^r) = r$, *if* $r > d(G)$ *and* G *is imperfect.*

Proof. Let $x_1,\dots,x_n$ be a coprime minimal set of generators of G, and set

$$x_{i,j} = (1,\dots,1, x_i, 1,\dots,1) \in G^r,$$

with 1's in all but the j-th position ($1 \le j \le r$). Assume that $r \le d(G)$. For each $i = 1,\dots,n$, define $g_i = x_{i,1}\, x_{i+1,2} \dots x_{i+r-1,r}$, where the indices are modulo n. By the coprimality hypothesis, one has that $x_{i+j-1,j} \in < g_i >$. Since $r \le d(G)$, every $x_{i,j}$ is a factor of some g_k. It follows that $G^r = < g_1,\dots,g_n >$. But obviously $d(G^r) \ge d(G) = n$. Hence $d(G^r) = d(G)$. Suppose next that $r > d(G)$

and G is imperfect. Since G/G' is a non-trivial cyclic group, from the natural epimorphism $G^r \to G^r/(G')^r$ one gets that $d(G^r) \geq r$. For each $i = 1,\ldots,r$, define

$$h_i = x_{1,i}\, x_{2,i+1} \ldots x_{n,i+n-1},$$

where the indices are modulo r. Then $x_{j,i+j-1} \in \langle h_i \rangle$, and since every $x_{i,j}$ is a factor of some h_k, we deduce that $G^r = \langle h_1,\ldots,h_r \rangle$. Thus $d(G^r) = r$.

4. Minimal number of generators of certain groups

In this section we shall describe bounds for the minimal number of generators of certain groups arising from wreath product constructions, with precise calculations in most cases.

Theorem 4.1. *Let* G *be a finitely generated group such that* $d(G/G') = d(G)$, *and let* C_m *be the cyclic group of order* $m \geq 2$. *Then* $d(G \wr C_m) = d(G) + 1$.

Proof. First note that we may assume that G is abelian by Lemma 3.1. Next observe that there exists a prime number p, and an epimorphism $G \to G_p^s$, where $s = d(G)$ (choose p to be a divisor of the order of the first group G_1 in the canonical decomposition of the abelian group G, with the notation of Section 3). Therefore, again by Lemma 3.1, we may assume that $G = C_p^s$, and by Lemma 3.2, we may suppose that $m = q$ is also a prime number. Obviously $d(G \wr C_q) \leq s+1$. We will prove the reverse inequality by showing that a subset of s elements cannot generate $G \wr C_q$. If $q=p$, this follows from Lemma 3.1. Assume then $p \neq q$. Write $G \wr C_q = K \rtimes C_q$, where $K = G^q = C_p^{sq}$ is a vector space of dimension sq over the field with p elements. Say that $C_q = \langle u \rangle$. Then every element of $G \wr C_q$ can be written uniquely in the form ku^i, with $k \in K$ and $0 \leq i \leq q-1$. Consider s elements $a_1,\ldots,a_s$ of $G \wr C_q$, and let $H = \langle a_1,\ldots,a_s \rangle$. If each a_i is in K, then clearly $H \neq G \wr C_q$. So we may assume that one of them, say $a_s \notin K$, and in fact $a_s = ku$ for some $k \in K$, renaming the generator of C_q if necessary. Note that for each $i \neq s$, one can make an appropriate choice of an exponent r(i), to have $a_i a_s^{r(i)} \in K$. Since

$$H = \langle a_1,\ldots,a_s \rangle = \langle a_1 a_s^{r(1)},\ldots,a_{s-1} a_s^{r(s-1)}, a_s \rangle,$$

we may assume that $a_1,\ldots,a_{s-1} \in K$ and $a_s = ku \notin K$. Note that if $1 \leq i \leq s-1$,

$$a_s a_i = kua_i = ka_i^{u^{-1}} u = a_i^{u^{-1}} a_s.$$

Consequently, every element of H can be written in the form ga_s^t, where

$$g \in \langle a_i^v \mid i = 1,\dots,s-1,\ v = u^j,\ j = 0,\dots,q-1 \rangle,\ 0 \le t \le qp-1.$$

So there are at most $p^{(s-1)q}$ elements of type g. On the other hand $ga_s^t \in K$ if and only if $a_s^t \in K$, i.e., if and only if $t = qi$ for some $i = 0,\dots,p-1$. Hence the number of elements of $K \cap H$ is at most $p^{(s-1)q+1}$. Since the order of K is p^{sq}, it follows that $K \cap H \neq K$, and thus $H \neq G \wr C_q$, as desired.

Corollary 4.2. *For any finite nilpotent group* G, $d(G \wr C_m) = d(G) + 1$, *if* $m \ge 2$.

Corollary 4.3. *If* G *is a finitely generated imperfect group, there exists a positive integer* k *such that* $d(G^r \wr C_m) = d(G^r) + 1$ *for all* $r \ge k$ *and* $m \ge 2$.

Proof. By Theorem 2.2 in [9], if $r \ge k = 2(2 + d(G)/d(G/G'))^2$, then

$$d(G^r) = r\,d(G/G') = d(G^r/(G^r)').$$

Therefore, if $r \ge k$, the conditions of Theorem 4.1 are satisfied.

Theorem 4.4. *Let* $C_n = \langle z \mid z^n = 1 \rangle$ *and* $D_m = \langle u,v \mid u^2 = v^m = (uv)^2 = 1 \rangle$ *be the cyclic group of order* n *and the dihedral group of order* 2m *respectively. Let* $G = C_n \wr D_m$. *Then* $d(G) = 3$ *if* $(m,n) \neq 1$, *and* $d(G) = 2$ *if* $(m,n) = 1$, *in fact in the latter case,* $G = \langle zu, zv \rangle$.

Proof. Obviously $2 \le d(G) \le 3$. Note that we have $G = C_n \wr D_m = K \rtimes D_m$, where K is a free $\mathbb{Z}/n\mathbb{Z}$-module of rank 2m. Let $a,b \in G$, and define $H = \langle a,b \rangle$, the subgroup generated by a and b. We shall show that if $(m,n) \neq 1$, then $H \neq G$. On the other hand if $(m,n) = 1$, we shall exhibit two specific elements a and b such that $H = G$.

Say $a = xu^iv^s$, $b = yu^tv^j$ ($i,t = 0,1$; $j,s = 0,\dots,m-1$; $x,y \in K$). Clearly, we may assume that either $i \neq 1$ or $t \neq 1$, for otherwise certainly $H \neq G$; and if $i,t = 1$, then, since $\langle a,ab \rangle = H$, we may suppose $t = 0$, i.e., $a = xuv^s$, $b = yv^j$. Now, if $s \neq 0$, then uv^s has order 2, and $\langle uv^s, v^j \rangle = D_m$ iff $(j,m) = 1$. Thus we may assume $j = 1$, and since $H = \langle ab^{-s},b \rangle$, we may assume $a = xu$ and $b = yv$.

Note

$$ab = xuyv = xy^uv^{-1}u = b^ux^{v^{-1}}u,$$

and inductively

$$ab^i = (b^u)^ix^{v^{-i}}u \text{ (for all } i \in \mathbb{Z}).$$

Similarly $a(b^u)^i = b^i(x^{v^i}u)$ (for all $i \in Z$). It follows that $H \leq B\bar{H}$, where $B = \langle b, b^u \rangle$, and $\bar{H} = \langle x^{v^i}u \mid i = 0,\ldots,m-1 \rangle$. Hence

$$H \cap K \leq B\bar{H} \cap K = (B \cap K)(\bar{H} \cap K).$$

Now,

$$B = \langle b, bb^u \rangle = \langle b, yvy^{uv^{m-1}} \rangle = \langle b, yy^{uv} \rangle,$$

and since $b^i yy^{uv} = y^{v^{-i}}y^{uv^{-i+1}}b^i$, it follows that

$$B = \langle y^{v^i}\, y^{uv^{i+1}} \mid i = 0,\ldots,m-1 \rangle \langle b \rangle.$$

Therefore,

$$\begin{aligned} B \cap K &= \langle y^{v^i}y^{uv^{i+1}}, b^m \mid i = 0,\ldots,m-1 \rangle \\ &= \langle y^{v^i}y^{uv^{i+1}}, yy^{v^{m-1}}y^{v^{m-2}} \ldots y^v \mid i = 0,\ldots, m-1 \rangle. \end{aligned}$$

Next note that $\bar{H} \cap K = \langle x^{v^i}\, x^{v^j u} \mid i,j = 0,\ldots, m-1 \rangle$. Thus $H \cap K$ is generated by the set

$$S = \{y^{v^i}\, y^{uv^{i+1}}, x^{v^i}\, x^{uv^j}, yy^{v^{m-1}}\, y^{v^{m-2}} \ldots y^v \mid i,j = 0, \ldots, m-1\}.$$

Now, K is a free module of rank 2m over the ring $\mathbf{Z}/n\mathbf{Z}$, and we can take

$$\mathfrak{B} = \{x_1, x_v, \ldots, x_{v^{m-1}}, x_{uv}, x_{uv^2}, \ldots, x_{uv^{m-1}}, x_u\}$$

as one of its ordered bases (written multiplicatively), where if z denotes a fixed generator for C_n, then $x_d := z^d = d^{-1}zd$, for every $d \in D_m$. Let us assume that the expressions of the elements x and y in terms of this basis are

$$x = \prod_{d \in D_m} x_d^{\alpha(d)}, \quad y = \prod_{d \in D_m} x_d^{\beta(d)}.$$

Write each element of S as a linear combination of the elements of $\mathfrak{B}$ with coefficients (i.e., exponents) in $\mathbf{Z}/n\mathbf{Z}$, and consider the matrix C whose rows are the ordered 2m-tuples of exponents corresponding to each element in S. Denote by $\bar{C}$ the submatrix of C consisting of all its rows except the one corresponding to the element $yy^{v^{m-1}}\, y^{v^{m-2}} \ldots y^v$. Note that the sum of the first m columns of $\bar{C}$ equals the sum of the last m columns of $\bar{C}$: for rows corresponding to elements of the form $y^{v^i}\, y^{uv^{i+1}}$, the sum of their first m entries is $\sum_{d \in D_m} \beta(d)$, and so is the sum of the last m entries; similarly for columns corresponding to elements of the form $x^{v^i} x^{uv^j}$, the sum of the first m entries is $\sum_{d \in D_m} \alpha(d)$, and so is the sum of the last m entries. On the other hand the sum of the first m entries of the row corresponding to $yy^{v^{m-1}} y^{v^{m-2}} \ldots y^v$ is

$$m(\beta(1) + \beta(v) + .. + \beta(v^{m-1})),$$

and the sum of its last m entries is $m(\beta(uv) + \ldots + \beta(uv^{m-1}) + \beta(u))$.

Assume now that $(m,n) \neq 1$. Let p be a prime number that divides (m,n). Then there is an epimorphism $C_n \wr D_m \to C_p \wr D_m$, by Lemma 3.1. Hence to prove that $d(C_n \wr D_m) = 3$, we may assume $n = p$, a prime number, such that $p \mid m$. Then the matrix C has entries in the field $\mathbf{Z}/p\mathbf{Z}$, and since $p \mid m$, we have

$$m(\beta(1) + \beta(v) + .. + \beta(v^{m-1})) = m(\beta(uv) + \ldots + \beta(uv^{m-1}) + \beta(u)) = 0.$$

Hence the sum of the first m columns of C equals the sum of its last m columns. Therefore the rank of C is at most 2m-1, and so the elements of S do not span the $\mathbf{Z}/p\mathbf{Z}$-vector space K, since K has dimension 2m. Thus $H \neq G$.

Now suppose that $(m,n) = 1$. Choose x and y to be the particular elements $x = x_1 = z$, and $y = x_1 = z$, i.e., with the above notation, $\alpha(1) = \beta(1) = 1$, and $\alpha(d) = \beta(d) = 0$ if $1 \neq d \in D_m$. Consider then the submatrix C' of C whose rows correspond to the 2m elements

$$S' = \{yy^{uv}, y^v y^{uv^2}, \ldots, y^{v^{m-1}} y^u, x\,x^u, x^v x^u, \ldots, x^{v^{m-2}} x^u, yy^{v^{m-1}} y^{v^{m-2}} \ldots y^v\}.$$

Then C' is the matrix

$$\left[\begin{array}{cccc|cccc}
1 & 0 & \ldots & 0 & 0 & 1 & 0 & \ldots & 0 & 0 \\
0 & 1 & \ldots & 0 & 0 & 0 & 1 & \ldots & 0 & 0 \\
\multicolumn{5}{c|}{\cdots\cdots\cdots} & \multicolumn{5}{c}{\cdots\cdots\cdots} \\
0 & 0 & \ldots & 0 & 1 & 0 & 0 & \ldots & 0 & 1 \\
\hline
1 & 0 & \ldots & 0 & 0 & 0 & 0 & \ldots & 0 & 1 \\
0 & 1 & \ldots & 0 & 0 & 0 & 0 & \ldots & 0 & 1 \\
\multicolumn{5}{c|}{\cdots\cdots\cdots} & \multicolumn{5}{c}{\cdots\cdots\cdots} \\
0 & 0 & \ldots & 1 & 0 & 0 & 0 & \ldots & 0 & 1 \\
1 & 1 & \ldots & 1 & 1 & 0 & 0 & \ldots & 0 & 0
\end{array}\right]$$

Note that $\det(C') = m$. Since $(n,m) = 1$, m is a unit in the ring $\mathbf{Z}/n\mathbf{Z}$, and therefore the elements in S' form a basis for the free $\mathbf{Z}/n\mathbf{Z}$-module K. In particular S' generates K, and consequently,

$$G = H = < a,b > = < xu, yv > = < x_1u, x_1v >;$$

i.e., if (m,n), $d(G) = 2$.

Corollary 4.5.

$d(D_m \wr D_n) = 4$ *if* m *and* n *are even,*

$d(D_m \wr D_n) = 3$ *if* m *and* n *have different parity, and*

$d(D_m \wr D_n) \leq 3$ *if* m *and* n *are odd.*

Proof. Say $D_m = < x,y \mid x^2, y^m, (xy)^2 >$ and $D_n = < u,v \mid u^2, v^n, (uv)^2 >$. Clearly $d(D^m \wr D_n) \leq 4$. If t is even, then $D_t/(D_t)' = C_2 \times C_2$. Suppose m and n are even; then by Lemma 3.1,

$$d(D_m \wr D_n) \geq d(D_m/(D_m)' \times D_n/(D_n)') = 4,$$

and hence $d(D_m \wr D_n) = 4$. Assume that n is even and m is odd; then by Lemma 3.1,

$$d(D_m \wr D_n) \geq d(D_m/(D_m)' \times D_n/(D_n)') = 3,$$

and by Proposition 3.4, $d(D_m \wr D_n) \leq 3$. If n is odd, then by Theorem 4.4 the subgroup $< x,u,v > \approx C_2 \wr D_n$ of $D_m \wr D_n$ can be generated by xu and xv, and so $D_m \wr D_n$ can be generated by xu, xv and y, i.e., $d(D_m \wr D_n) \leq 3$; on the other hand, if in addition m is even $d(D_m \wr D_n) \geq 3$, by Lemma 3.1, and therefore, $d(D_m \wr D_n) = 3$.

Corollary 4.6. *Let* G *be a finitely generated imperfect group which admits a coprime minimal set of generators* $x_1,\ldots,x_t$ (see Section 3). *Let* m *be a natural number not coprime with the order of at least one of the generators* x_i, *and assume that* $t \leq 2m$. *Then* $d(G \wr D_m) = 3$.

Proof. By Proposition 3.4, $d(G \wr D_m) \leq 3$. Since G/G' is a finite cyclic group of order not coprime with m, it follows from Theorem 4.4 that $d(G/G' \wr D_m) = 3$, and therefore by Lemma 3.1, $d(G \wr D_m) \geq 3$.

Corollary 4.7. *Let* m *be an even integer and consider the dicyclic group of order* $2m$,

$$Q_m = < u, v \mid v^m = 1, v^{m/2} = u^2, u^{-1}vu = v^{-1} >.$$

Then $d(C_n \wr Q_m) = 3$ *if* $(n, m/2) \neq 1$.

Proof. Certainly $d(C_n \wr Q_m) \leq 3$. Clearly the dihedral group $D_{m/2}$ is a homomorphic image of Q_m (simply add the relation $u^2 = 1$). So by Lemma 3.2, $d(C_n \wr Q_m) \geq d(C_n \wr D_{m/2})$. Finally, since $(n, m/2) \neq 1$, $d(C_n \wr D_{m/2}) = 3$ by Theorem 4.4.

Using Lemma 3.1 one gets, as a consequence, the following special cases of a result of Gruenberg (cf. [3, Ths. 6.1 and 6.9]).

Corollary 4.8. $d(\mathbf{Z} \wr D_m) = 3$ *and* $d(\mathbf{Z} \wr Q_m) = 3$.

Corollary 4.9. *If* $n/2$ *and* $m/2$ *are even, then* $d(Q_n \wr Q_m) = 4$; *and if* $n/2$ *is odd and* $m/2$ *is even, then* $d(Q_n \wr Q_m) = 3$.

Proof. Note that if t is even, then $Q_t/(Q_t)' \approx C_2 \times C_2$. Therefore if both $n/2$ and $m/2$ are even, there is an epimorphism (cf. Lemma 3.1)

$$Q_n \wr Q_m \to Q_t/(Q_t)' \times Q_t/(Q_t)' \approx C_2 \times C_2 \times C_2 \times C_2\,;$$

and thus $d(Q_n \wr Q_m) = 4$. Suppose now that $n/2$ is odd and $m/2$ is even. Then $Q_n \approx C_{n/2} \rtimes C_4$, i.e., has a coprime minimal set of generators. Then by Proposition 3.4, $d(Q_n \wr Q_m) \leq 3$. On the other hand, there exists an epimorphism

$$Q_n \wr Q_m \to C_4 \times Q_t(Q_t)' \approx C_4 \times C_2 \times C_2,$$

and so $d(Q_n \wr Q_m) \geq 3$.

Acknowledgements

The first author was partially supported by NSERC Grant OGP0008079.

We thank M Newman and A Lubotzky for useful conversations on the topics of this paper.

References

1. N Bourbaki, *Topologie Générale, Ch. 1* (3ième ed., Hermann 1961).
2. I Grushko, Über die Basen eines freien Produktes von Gruppen, *Mat. Sb.* **8** (50) (1940), 19-182.
3. K W Gruenberg, *Relation modules of finite groups* (Reg. Conf. Ser. Math., No. 25, Amer. Math. Soc., 1974).
4. W N Herfort & L Ribes, Subgroups of free pro-p-products, *Math. Proc. Cambridge Philos. Soc.* **101** (1987), 197-206.
5. A Lubotzky, Combinatorial group theory for pro-p-groups, *J. Pure Appl. Algebra* **25** (1982), 311-325.
6. B H Neumann, On the number of generators of a free product, *J. London Math. Soc.* **18** (1943), 12-20.
7. L Ribes, *Introduction to profinite groups and Galois cohomology* (Queen's Papers in Pure and Appl. Math., No. 24, Kingston, 1970).

8. J-P Serre, *Cohomologie Galoisienne* (Lect. Notes in Math. **5**, Springer 1965).

9. J Wiegold & J S Wilson, Growth sequences of finitely generated groups, *Arch. Math.* **30** (1978), 337-343.

LIE PROPERTIES OF MODULAR GROUP ALGEBRAS

ELIYAHU RIPS & ANER SHALEV

Hebrew University of Jerusalem, Jerusalem 91904, Israel

Abstract

A Lie algebra is said to be BB (Baer-bounded) if all its 1-dimensional sub-algebras are subnormal of bounded index. We give a complete characterization of group algebras having this property (as Lie algebras). In particular we prove: (I) every BB group algebra of characteristic $p > 2$ is Lie-nilpotent, and (II) every BB group algebra of characteristic 2 is Lie-soluble. Using groups of units, we then construct a locally nilpotent n-Engel group which is not a Baer group, thus providing an answer to a question of Plotkin.

1. The Engel condition

Let $p > 0$ be a fixed prime, and let K be a field of characteristic p. If G is any group, consider the modular group algebra KG, regarded as a Lie algebra (with the usual bracket operation). Its properties were studied by Passi, Passman and Sehgal in the early seventies.

Theorem 1.1 (Passi, Passman, Sehgal [PPS]).

(a) KG *is Lie-nilpotent iff* G *is nilpotent and p-abelian (i.e. the commutator subgroup* G′ *is a finite p-group).*

(b) *Suppose* $p > 2$. *Then* KG *is Lie-soluble iff* G *is p-abelian.*

(c) *Suppose* $p = 2$. *Then* KG *is Lie-soluble iff* G *possesses a 2-abelian subgroup of index at most 2.*

Another important Lie-property is the Engel property. A Lie algebra L is said to be n-*Engel* if it satisfies the identity

$$[x,\underbrace{y,y,\dots,y}_{n}] = 0.$$

An n-Engel group algebra of characteristic 0 is easily seen to be commutative. A group theoretical characterization of n-Engel group algebras of characteristic p is given by:

Theorem 1.2 (Sehgal [Se, Chap. V, Sec. 6]). KG *is* n-*Engel for some* n *iff* G *is a nilpotent group possessing a p-abelian subgroup of finite index, and* G/Z(G) *is a p-group of finite exponent.*

Comparing Theorem 1.1 and Theorem 1.2, it is evident that the Engel condition does not imply Lie-nilpotency (or even Lie-solubility) in modular group algebras. It turns out that the situation is different when KG satisfies the n-Engel condition with small n.

Proposition 1.3. *Suppose* KG *is* n-*Engel:*

(a) *If* $n < p$, *then* KG *is commutative.*

(b) *If* $n = p$, *then* $|G'| \in \{1,p\}$; *consequently* KG *is Lie-nilpotent of index* p+1 *at most (the converse is trivial).*

(c) *If* $n \leq 2p-2$, *then* KG *is Lie-nilpotent.*

(d) *If* $p = 2$ *and* $n \leq 6$, *then* KG *is Lie-soluble.*

The proof of these observations, as well as of the main results stated below, will be given in [RS].

It should be noted that, although n-Engel group algebras need not be Lie-nilpotent, they turn out to be residually Lie-nilpotent. This may be verified directly, using the characterization of residually-Lie-nilpotent group rings, given by Passi in [Pa, Chap. VI, Thm. 2.28]. It would be interesting to construct an n-Engel Lie algebra which is not residually nilpotent (if such exists).

2. The Baer Condition

In 1986 Zel'manov proved the remarkable result, that every n-Engel Lie algebra over a field of characteristic 0 is nilpotent [Ze]. As we have already seen, the situation is quite different in positive characteristic. Constructions of non-nilpotent (and non-soluble) n-Engel Lie algebras of characteristic p were provided long ago by Cohn [Co] and Razmyslov [Ra]. Theorems 1.1 & 1.2 enable one to provide simpler examples, consisting of modular group algebras.

For a long time it was not known whether every n-Engel Lie algebra of characteristic $p > 0$ is locally nilpotent, as states the celebrated theorem of Kostrikin [Ko1, Ko2] in the case $n \leq p$ (see also [Br] for $n = p+1$). Recently it was announced that this has been settled in the affirmative by Zel'manov, in his solution to the restricted Burnside problem. However, it seems natural to ask what stronger assumption on non-finitely-generated Lie algebras of positive characteristic would imply nilpotency.

A natural candidate is the Baer condition. A Lie algebra is said to be BB (Baer-bounded) if all its 1-dimensional sub-algebras are subnormal of bounded index. This is equivalent to: there exists a positive integer n, such that, for every x in L, there exists a series

$$L = L_0 \supseteq L_1 \supseteq \ldots \supseteq L_n = 0$$

of Lie sub-algebras, satisfying

$$[L_{i+1}, L_i] \subseteq L_{i+1} \text{ and } [x, L_i] \subseteq L_{i+1} \ (0 \le i < n).$$

If this holds, we say that L is n-*Baer*. It turns out that, when dealing with group algebras, this condition is considerably strong.

Our main result is:

Theorem 2.1. *Every* BB *group algebra of characteristic* $p > 2$ *is Lie-nilpotent.*

The case $p = 2$ is more complicated: we show that there exists a 3-Baer group algebra KG of characteristic 2, which is not Lie-nilpotent. However, it is possible to prove the following result:

Theorem 2.2. *Every* BB *group algebra of characteristic* 2 *is Lie-soluble.*

In fact, we show that a group algebra of characteristic 2 is BB iff it is Lie-soluble and n-Engel for some n.

These results, combined with previous characterizations, provide a complete group-theoretical characterization of the groups G for which KG is Baer-bounded (where K is any field).

3. On a question of Plotkin

Regarding Theorems 2.1 and 2.2, one may ask whether every BB Lie algebra is soluble. It turns out that the answer is 'no', even for Lie algebras arising from associative rings. The following example of a non-soluble BB Lie algebra also gives rise to some interesting group theoretical constructions.

We first revise some notions. A group G is called *bounded left Engel*, if for every $y \in G$, there exists a positive integer $n = n(y)$, such that $(x,\underbrace{y,y,\ldots,y}_{n}) = 1$ for all $x \in G$, where (x,y) denotes a group commutator. If $n(y) \le n$ for all y, we say that G is n-Engel. Finally, a Baer group is a group in which every element is subnormal.

Theorem 3.1. *Let* $p > 2$, *and let* $G = A \rtimes \langle x \rangle$, *where* $A = \langle y_i, z_i \rangle$ *is an infinite elementary abelian* p*-group, and* x *is an element of order* p, *acting on* A *by*

$$y_i^x = y_i \cdot z_i, z_i^x = z_i \; (i < \omega).$$

For a field K *of characteristic* p, *denote the augmentation ideal of* KA *by* $\Delta(A)$, *and consider the associative ring (without a unit)* $R = KG \cdot \Delta(A)$ *(regarded as a subring of KG). Set* $n = 1 + (p-1)p^2$. *Then:*

(a) R *is a non-soluble* n*-Baer Lie algebra.*

(b) *The multiplicative group* $H = 1+R$ *is a non-soluble* p*-group of exponent* $\leq p^3$, *which is a union of normal nilpotent subgroups of class less than* n.

(c) *The wreath product* $M = H \text{ wr } C_p$ *is a locally nilpotent* k*-Engel group for some* k, *which is not a Baer group.*

For part (a) we use the fact that $(KGyKG)^n = 0$ for all y in R. The non-solubility of H in part (b) follows from the non-solubility of U(KG) - the group of units of KG (see [Ba] and [BK] for a complete characterization of group algebras with a soluble group of units). Part (c) is then established using the above-mentioned properties of H.

This construction settles a problem raised by Plotkin, on the existence of a locally nilpotent bounded left Engel group, which is not a Baer group (see [Pl, Chap. V, §4, Sec. 2]; see also [Ku, §D.26, Sec. 4]). We note that the non-soluble group built by Razmyslov in [Ra] may be used instead of H in the construction of M.

References

[Ba] J M Bateman, On the solvability of unit groups of group algebras, *Trans. Amer. Math. Soc.* **157** (1971), 73-86.

[BK] A A Bovdi & I I Khripta, Group algebras of torsion groups with a soluble multiplicative group, *Mat. Zametki* **22** (1977), 421-432.

[Br] A Braun, Lie rings and the Engel condition, *J. Algebra* **31** (1974), 287-292.

[Co] P M Cohn, A non-nilpotent Lie ring satisfying the Engel condition and a non-nilpotent Engel group, *Proc. Cambridge Philos. Soc.* **51** (1955), 401-405.

[Ko1] A I Kostrikin, On the Burnside problem, *Izv. Akad. Nauk. SSSR, Ser. Mat.* **23** (1959), 3-34.

[Ko2] A I Kostrikin, *Around Burnside* (Nauka, Moscow, 1986).

[Ku] A G Kurosch, *Theory of Groups*, 3rd edition (Nauka, Moscow, 1967).

[Pa] I B S Passi, *Group Rings and their Augmentation Ideals* (Lecture Notes in Math. **715**, Springer-Verlag, Berlin-Heidelberg-New York, 1979).

[PPS] I B S Passi, D S Passman & S K Sehgal, Lie solvable group rings, *Canad. J. Math.* **25** (1973), 748-757.

[Pl] B I Plotkin, *Groups of Automorphisms of Algebraic Systems* (Walters-Noordhoff Publ., Groningen, 1972), English translation.

[Ra] Y P Razmyslov, On Lie algebras satisfying the Engel condition, *Algebra and Logic* **10** (1971), 21-29.

[Ri] E Rips, Iteration of operations on classes of groups, *J. Algebra* **80** (1983), 37-59.

[RS] E Rips & A Shalev, The Baer condition for group algebras, *J. Algebra*, to appear.

[Se] S K Sehgal, *Topics in Group Rings* (Marcel Dekker, New York-Basel, 1978).

[Sh] A Shalev, Lie dimension subgroups, Lie-nilpotency indices, and the exponent of the group of normalized units, *J. London Math. Soc.*, to appear.

[Ze] E I Zel'manov, On Engel Lie algebras, *Siberian Math. J.* **29** No. 5 (1988), 112-117.

OBSERVATIONS ON A CONJECTURE OF HANS ZASSENHAUS

KLAUS W ROGGENKAMP

Math. Institut B, Universität Stuttgart, West Germany

This is a report on joint work with Leonard L Scott [RS1].

1. Introduction

Let G be a finite group and R an integral domain of characteristic zero, in which no rational prime divisor of $|G|$ is invertible, K is the field of fractions of R. By RG we denote the group ring of G over R, and $\varepsilon : RG \to R$ is the augmentation.

The following is a **conjecture of Hans Zassenhaus** for $R = \mathbf{Z}$:

(1.1) *Let* RG = RH *as augmented algebras for a finite group* H. *Then there exists a unit* $a \in$ KG *such that* $aHa^{-1} = G$. *(Such an element a would then automatically normalize* RG.)

Since the hypotheses on R guarantee that finite subgroups in V(RG), the units of augmentation one in RG, are linearly independent over R [B], it is enough to assume in (1.1), that H is a finite subgroup in V(RG) with $|G| = |H|$.

The **isomorphism problem** asks,

(1.2) whether RG = RH implies that G and H are isomorphic.

A positive answer to the Zassenhaus conjecture would settle the isomorphism problem positively, but it also would give information about the embedding of H in V(RG).

If the isomorphism problem has a positive answer, then the Zassenhaus conjecture is equivalent to :

(1.3) *Let* α *be an augmentation preserving automorphism of* RG. *Then* α *is the composition of an automorphism induced from a group automorphism followed by a central automorphism; i.e., an automorphism fixing the centre elementwise. (This is just the Skolem-Noether theorem.)*

Then the centre of RG is generated over R by the class sums $\{K_g\}$, where

$$K_g = \sum_{x \in G/C_G(g)} {}^x g, \quad ({}^x g = x \cdot g \cdot x^{-1}).$$

A result of Berman [B] - extended in [RS] - states:

(1.4) *If* RG = RH *as augmented algebras, then the class sums of* G *and of* H *are the same in* RG.

Here is another interpretation of the Zassenhaus conjecture:

(1.5) *Let* H *be a finite subgroup of* V(RG) *with* $|G| = |H|$. *Then there exists an isomorphism* $\rho : G \to H$ *with* $K_{\rho(g)} = K_g$ *in* RG = RH *for every* $g \in G$, *if and only if the Zassenhaus conjecture holds for* RG.

Proof. Since $\hat{\rho}$, the automorphism of RG induced from ρ, fixes the centre elementwise, it is given by conjugation with an element $a \in KG$. Conversely, if the Zassenhaus conjecture holds, then the conjugation with $a \in KG$ gives an isomorphism from G to H centralizing the class sums.

In terms of automorphisms this can be phrased as follows:

(1.6) *Let* $\alpha : RG \to RG$ *be an automorphism commuting with the augmentation. Then the Zassenhaus conjecture holds for* α *if and only if there exists a group automorphism* ρ *of* G *such that for every* $g \in G$, $K_{\alpha(g)} = K_{\rho(g)}$.

With other words, the action of any augmentation preserving automorphism ρ of RG on the centre of RG can be compensated therefore by a group automorphism of G. If one observes that the centre of **Z**G is a **Z**-order in $\Pi\, S_i$, where S_i are rings of algebraic integers in algebraic number fields, then on the centre, one might have **Z**-automorphisms of S_i combined with permuting the central idempotents. The Zassenjaus conjecture now states that all these can be compensated for by group automorphisms. This is a very strong statement, and it is somewhat surprising that there are positive results:

(1.7) *The Zassenhaus conjecture is true for* R = **Z** *if*

(a) G *is nilpotent* [RS],

(b) G *is such that the generalized Fitting subgroup is a* p-*group* [RS2,S] - *in the solvable case, these are groups* G *such that there exists a rational prime* p *with* $O_{p'}(G) = 1$, *where* $O_{p'}(G)$ *is the largest normal subgroup of* G *of order prime to* p.

(1.7) covers all the previously known results cf. [San].

We have used (1.7(a)) to show that

(1.8) *the isomorphism problem has a positive answer, if the finite group* G *has a normal abelian subgroup* A *such that* G/A *is nilpotent.*

We tried hard to prove the Zassenhaus conjecture for such groups - however, we were not successful.

On the other hand, every finite solvable group G is built up - in a reasonably understood way as a subdirect product - from the various $G/O_{p'}(G)$ (here $O_{p'}(G)$ is the largest normal subgroup of G of order prime to p); i.e., from groups where the Zassenhaus conjecture is true. We have tried to analyse the underlying patchwork process, and we found a serious obstruction for the validity of the Zassenhaus conjecture for solvable groups.

In fact we found a metabelian group, which is a counterexample to the Zassenhaus conjecture.

It should be noted, that the isomorphism problem has a positive answer for metabelian groups [W, J, (1.8)]. In view of (1.7(b)) this substantially increases the chances that the isomorphism problem itself will eventually have a negative answer, perhaps for a group not too much more complicated than the one we have constructed. The example, which I shall describe, is not a random event, but the result of systematic investigations of possible automorphisms of group rings.

2. The group of the counterexample and philosophy

We shall first describe the group G of the counterexample - note that this example stands for a whole family of groups, and we have just chosen the one of smallest order:

Let $F_4 = \{0,1,\xi,\xi^2\}$ be the field with 4 elements. In the ring of (3×3)-matrices over F_4 we consider the following matrices:

$$s = \begin{bmatrix} 1 & 1 & 0 \\ 0 & 1 & 0 \\ 0 & 0 & 1 \end{bmatrix}, \quad t = \begin{bmatrix} 1 & 0 & 0 \\ 0 & 1 & 1 \\ 0 & 0 & 1 \end{bmatrix},$$

$$u = \begin{bmatrix} 1 & \zeta & 0 \\ 0 & 1 & 0 \\ 0 & 0 & 1 \end{bmatrix}, \quad v = \begin{bmatrix} 1 & 0 & 0 \\ 0 & 1 & \zeta \\ 0 & 0 & 1 \end{bmatrix}, \tag{2.1}$$

and also

$$c = \begin{bmatrix} 1 & 0 & 1 \\ 0 & 1 & 0 \\ 0 & 0 & 1 \end{bmatrix}, \quad c' = \begin{bmatrix} 1 & 0 & \zeta \\ 0 & 1 & 0 \\ 0 & 0 & 1 \end{bmatrix}, \quad c'' = \begin{bmatrix} 1 & 0 & \zeta^2 \\ 0 & 1 & 0 \\ 0 & 0 & 1 \end{bmatrix}.$$

(2.2) *The elements* s,t,u,v *generate the group of order* 4^3, *which we denote by* H_o, *of* 3×3 *upper triangular matrices over* F_4 *with diagonal entries* 1. *We put* $Z = \{1,c,c',c''\}$, *the centre of* H_o.

Recall that an automorphism of a group is *central* if it stabilizes all of the conjugacy classes.

(2.3) H_o *has a central automorphism* σ_o, *which is not an inner automorphism, and which is given on generators as follows:*

$$\sigma_o(s) = cs \quad \text{and} \quad \sigma_o(t) = ct,$$

$$\sigma_o(u) = u \quad \text{and} \quad \sigma_o(v) = v.$$

The fixed group of σ_o is $< st,u,v,Z >$.

In order to define our group G we let

$$M = < m: m^3 = 1 >,$$

$$N = < n: n^3 = 1 > \text{ and}$$

$$Q = < q: q^5 = 1 >.$$

We now define an action of H_o on $M \times N \times Q$:

$$sms^{-1} = m^{-1},\ tmt^{-1} = m^{-1}, \text{ and all other given generators of } H_o \text{ fix m.} \quad (2.4)$$

$$tnt^{-1} = n^{-1}, \text{ and all other given generators of } H_o \text{ fix n.}$$

$$sqs^{-1} = q^{-1}, \text{ and all other generators of } H_0 \text{ fix q.}$$

Definition 2.5. We let H be the semidirect product

$$H = M \cdot H_o,$$

and we put

$$G = (N \times Q) \cdot H,$$

$$G_3 = N \cdot H \text{ and } G_5 = Q \cdot H.$$

(2.6) The automorphism σ_o of H_o extends to an automorphism σ of H, which is central on H with $\sigma(m) = m$, and to automorphisms σ_3 of G_3, with

$$\sigma_3(x \cdot n) = \sigma(x) \cdot n,$$

σ_5 of G_5, with

$$\sigma_5(x \cdot q) = \sigma(x) \cdot q,\ x \in H,$$

and hence to σ_G of G. However, none of the extensions to G_3, G_5, or G can be chosen to be central.

The crucial property of G as a candidate for a counterexample to the Zassenhaus conjecture is the following:

(2.7) *Let* $\rho_3 : G_3 \to G_3$ *and* $\rho_5 : G_5 \to G_5$ be **central group automorphisms.** *Then we always have*

$$\bar{\rho}_3 \neq \bar{\rho}_5\, \sigma,$$

where $\bar{\rho}_3$ *and* $\bar{\rho}_5$ *are the corresponding automorphisms induced modulo* N *and* Q, *respectively, by the above maps*. (Notice $\rho_3(N) = N$ and $\rho_5(Q) = Q$, since ρ_3, ρ_5 are central.)

Finally, we note a property of the group G which is critical for the global arguments later on:

(2.8) *The group* G *satisfies the Eichler condition.*

Our group G is a pull-back

$$\begin{array}{ccc} G & \to & G_5 \\ \downarrow & & \downarrow \\ G_3 & \to & H \end{array}, \qquad (2.8)'$$

where the maps to H are surjective.

It is by no means the case that we can write a corresponding pull-back diagram for the group rings, but it is, nevertheless, instructive to form a pull-back diagram of rings

$$\begin{array}{ccc} \Gamma & \to & \mathbf{Z}G_5 \\ \downarrow & & \downarrow \\ \mathbf{Z}G_3 & \to & \mathbf{Z}H \end{array}, \qquad (2.9)$$

where Γ is just the **Z**-order defined by the pull-back. The order Γ is only a relatively small homomorphic image of the group ring **Z**G. Nevertheless, we may regard it as a first approximation for our analysis.

Before going into any further details of the proof of the counterexample, it is best first to understand more fully what the problem is.

(2.10) **The obstruction**: Suppose we have an automorphism α of ZG which behaves well with respect to the projections on $\mathbf{Z}G_3$ and $\mathbf{Z}G_5$, inducing automorphisms α_3 and α_5, respectively. (The automorphism α must also then stabilize the intersection and sum of the kernels of these projections, hence induces automorphisms α_Γ and α_{ZH} of Γ and ZH, respectively.)

Let us assume that α_3 and α_5 are both compatible with the Zassenhaus conjecture, so that we may write

$$\alpha_3 = \beta_3\, \delta_3,$$

$$\alpha_5 = \beta_5\, \delta_5,$$

where β_i is a group automorphism of ZG_i and δ_i is a central automorphism, i = 3 or 5. Such products henceforth will be called *Zassenhaus factorizations.* To simplify notation, we will use the same symbol for automorphisms of groups and the automorphisms they induce on their group rings.

Observe that the central automorphisms δ_i induce automorphisms of ZH, from which it follows that each β_i does also, and, consequently, β_i induces an automorphism of the group H.

Consider now the automorphism $\sigma = \beta_5^{-1}\, \beta_3$ *of the group* H. *This automorphism is central* in the group-theoretic sense. That is, it stabilizes all the conjugacy classes of H. (On H both of the expressions for α_i above are equal, giving $\sigma = \beta_5^{-1}\, \beta_3 = \delta_5\delta_3^{-1}$, which fixes the class sums.) Suppose α is compatible with the Zassenhaus conjecture, so that it has a Zassenhaus factorization $\alpha = \beta\delta$, where β is a group automorphism of ZG and δ is a central automorphism. Then $\beta^{-1}\beta_3 = \delta\delta_3^{-1}$ is a central group automorphism on G_3, and $\beta^{-1}\beta_5$ is a central group automorphism on G_5. Put $\rho_3 = \beta^{-1}\beta_3$ and $\rho_5 = \beta^{-1}\beta_5$. Each of the central group automorphisms ρ_3, ρ_5 induces an automorphism of the group H, and on H we have

$$\sigma = \beta_5^{-1}\beta_3 = \rho_5^{-1}\rho_3.$$

To summarize:

(2.11) *If* α *is compatible with the Zassenhaus conjecture, then the central group automorphism* α *of* H *must be the product of automorphisms of* H *induced by central group automorphisms of* G_3 *and* G_5.

Write $\mathrm{Aut_c(H)}$ for the group of central automorphisms of any given group H, and, for a central automorphism ρ of a group having H as a quotient, let ρ or $\bar{\rho}$ or $\rho|_H$ denote the automorphism ρ induces on H. For σ in $\mathrm{Aut_c(H)}$, with H as above, let $[\sigma]$ denote the double coset of the subgroup pair

$\mathrm{Aut_c}(G_5)|_H, \mathrm{Aut_c}(G_3)|_H$

to which σ belongs. Then it is straightforward to check, for $\sigma = \beta_5^{-1}\beta_3$ as above, that $[\sigma]$ depends only on the automorphisms α_3 and α_5, and not on the choice of Zassenhaus factorizations for these two automorphisms. In particular, $[\sigma]$ *depends only on* α, *even only on* α_Γ, *and we write* obs(α) = obs(α_Γ) = $[\sigma]$. It does, in fact, make sense *to talk about a given automorphism* α_Γ *having a Zassenhaus factorization.* This is just a product $\alpha_\Gamma = \beta\delta$ where β is an automorphism of Γ stabilizing the evident copy of G in Γ (equivalently, β is induced by a group automorphism of **Z**G), and δ is a central ring automorphism of Γ. Also, obs(α_Γ) makes sense, whether α_Γ is induced by an automorphism of **Z**G or not, and we have:

(2.12) *Suppose* α_Γ *is an automorphism of* Γ *inducing automorphisms* α_3 *and* α_5 *of* $\mathbf{Z}G_3$ *and* $\mathbf{Z}G_5$, *respectively, each with Zassenhaus factorizations. Then* α_Γ *has a Zassenhaus factorization if and only if* obs(α_Γ) = [1]. *If* α_Γ *is, in fact, induced by an automorphism* α *of* **Z**G, *then* obs(α_Γ) = obs(α), *which must be* [1] *if* α *has a Zassenhaus factorization.*

This is just a restatement of the above discussion, excepting the converse for α_Γ, which is straightforward to check. It may be noticed that the double coset space above is formally equivalent to a Cech 1-cohomology group for a two element open cover of a topological space. Bass [B,XIV,§5], [B2] uses a similar principle in conceptualizing the vector bundle foundations of algebraic K-theory. The bundle point of view is quite helpful in attaching some meaning to the obstructions above.

Next, it is worth noting that the entire obstruction discussion above could be carried out with **Z** replaced by more general domains R, such as those described at the beginning of this paper, replacing Γ by $R\Gamma = R \otimes \Gamma$. Clearly, we have:

(2.13) *The analogue of* (2.12) *above holds for* R-*algebra automorphisms* $\alpha_{R\Gamma}$ *of* $R\Gamma$. *Moreover, if* $\alpha_{R\Gamma}$ *is the extension to* $R\Gamma$ *of an automorphism* α_Γ *satisfying the hypothesis of* (2.12), *then* obs($\alpha_{R\Gamma}$) = obs(α_Γ).

The obstructions above are also robust with respect to minor changes in the automorphism. We can multiply $\alpha_{R\Gamma}$ on the right (or left) by a central automorphism of $R\Gamma$ and still get the same obstruction. Indeed, obs($\alpha_{R\Gamma}$) just depends on the group automorphisms β_3 and β_5 appearing in the Zassenhaus factorization of $\alpha_{R\Gamma}$ on RG_3 and RG_5, respectively. Thus any automorphism of $R\Gamma$ which differs from $\alpha_{R\Gamma}$ on each of RG_3 and RG_5 by an inner automorphism, of these respective rings, gives the same obstruction. An important special case is:

(2.14) Obs($\alpha_{R\Gamma}$) *depends only on the image of* $\alpha_{R\Gamma}$ *in the outer automorphism group* Out($R\Gamma$) *of* $R\Gamma$, *and even only on the images of* $\alpha_{R\Gamma}$ *in* Out(RG_3) *and* Out(RG_5). (That is, any other automorphism of $R\Gamma$ with the same images in these outer automorphism groups has the same obstruction as $\alpha_{R\Gamma}$.)

We have now discussed everything we need for our counterexample.

Thus, at the purely group-theoretic level, we have a candidate obstruction [σ] by (2.6) and (2.7). The construction of an automorphism α of **Z**G for which it is realized is done very slowly. First, we construct a semilocal automorphism: *We use for the ring* R *the semilocalization* $\mathbf{Z}_\pi$ of **Z** at the set of primes consisting of 2, 3 and 5. (These primes will be just those that divide the group order |G|. By a semilocalization at π, we just mean the intersection of the localizations at the primes in π. A more general semilocal ring R than $\mathbf{Z}_\pi$ is used for the notation and arguments in Section 4, but R = $\mathbf{Z}_\pi$ is the main case of interest.) We show first that [σ] is the obstruction for an automorphism of $R\Gamma$, and then lift the automorphism to RG. We then repeat the process with R replaced by **Z**. In passing from the semilocal automorphisms to global versions, we encounter new possible obstructions in the projective class group.

3. The construction of a semilocal automorphism

First some general phenomena, which can largely be found in [RMO, §37]; cf. also [RS, §1]. Let R be a Dedekind domain with quotient field K, which we assume for simplicity has characteristic 0. Let Ω be an R-order in a semisimple K-algebra.

For any R-automorphism α of Ω we denote by

$$1^{\Omega}\alpha \tag{3.1}$$

the Ω-bimodule which is Ω as left module, but with the right action twisted by α:

$$\lambda\cdot x\cdot\lambda' = \lambda x\alpha(\lambda'), \qquad x, \lambda, \lambda' \in \Omega.$$

Elementary arguments show that automorphisms which differ only by multiplication by an inner automorphism lead to isomorphic bimodules, and conversely [RMO, (37.14)]. Also, it is clear that the bimodule ${}_1\Omega_\alpha$ is invertible [RMO, (37.12),(37.13)]. (The inverse is ${}_\alpha\Omega_1 \equiv {}_1\Omega_{(\alpha^{-1})}$.) It is not difficult to give a characterization of these bimodules [RS, (1.25)]: - (Cf. also [RMO, (37.16)], together with [RMO, Exercise 38.4].)

(3.2) *Let* M *be a bimodule for* Ω *over* R. *(We assume* R *acts the same on both sides.) Then*

$$M \simeq {}_1\Omega_\alpha \text{ as } \Omega \text{ bimodule,}$$

for some R*-automorphism* α *of* RG *if and only if*

$$M \simeq \Omega \text{ as left } RG\text{-module.}$$

We will not repeat the proof here, but it is worth recording that a choice of a left Ω-module isomorphism $M \simeq \Omega$ determines a unique automorphism α of Ω for which the given isomorphism of left modules becomes a bimodule isomorphism $M \simeq {}_1\Omega_\alpha$. (If the bimodule structure of M is transported to Ω using the given isomorphism, each element $x \in \Omega$ acts via right multiplication by a uniquely determined element $\alpha(x) \in \Omega$.) The converse is true as well, and obvious: Given an automorphism α of Ω and a bimodule isomorphism $M \simeq {}_1\Omega_\alpha$, that bimodule isomorphism is indeed a left module isomorphism, and the given automorphism α and bimodule isomorphism are recovered from the procedure above.

If R is semilocal - i.e., R has only finitely many maximal ideals - then the existence of an isomorphism of M with Ω can be checked "a prime at a time", at each localization (or completion) of R [RMO, (3.16) and (18.2) iii]. Thus we conclude:

(3.3) *If* R *is semilocal and* M *is a bimodule for* Ω *over* R *(that is,* R *acts the same on both sides), then*

$$M \simeq {}_1\Omega_\alpha \text{ as } \Omega\text{-bimodules}$$

for some automorphism α *of* Ω *iff for each maximal ideal* $\mathfrak{p}$ *of* R *there is an automorphism* $\alpha_\mathfrak{p}$ *such that*

$$M_\mathfrak{p} \simeq {}_1(\Omega_\mathfrak{p})_{\alpha_\mathfrak{p}} \text{ as } \Omega_\mathfrak{p}\text{-bimodules.}$$

Any such $\alpha_\mathfrak{p}$ *differs from the automorphism induced by* α *on* $M_\mathfrak{p}$ *at most by multiplication by an inner automorphism of* $\Omega_\mathfrak{p}$.

Here R_p could either be the localization or completion of R at p. By way of notation we will generally use R_p for the completion, reserving $R_{(p)}$ for the localization.

If the bimodule M above is not given, one can ask if it can be constructed from the local data. This is possible if and only if the bimodules K_pM_p all arise from a common $K\Omega$ module [RMO, (4.2)]. A natural candidate is $K\Omega$ itself.

This leads to the notion of a central bimodule: Returning to our general set-up, we say then an Ω-bimodule is *central* if the centre $Z(\Omega)$ acts the same on both sides. If M is also invertible, this implies $KM \simeq K\Omega$ [RMO, (37.21)]. An automorphism of Ω is called central if it fixes the centre. Observe that (3.1) and (3.2) hold for central bimodules and central automorphisms.

The group of isomorphism classes of invertible central Ω-bimodules is denoted $\text{Picent}(\Omega)$, and the group of central automorphisms modulo inner automorphisms is denoted $\text{Outcent}(\Omega)$. The construction (3.2) gives a natural inclusion

$$\text{Outcent}(\Omega) \subseteq \text{Picent}(\Omega).$$

For group rings these coincide if R is local (or semilocal) [RS, 1.2.12]. For general orders Ω as above an intermediate subgroup $\text{LFP}(\Omega)$ has been introduced [RMO, p.345], consisting of those classes [M] of central bimodules for which M_p is free on the left. Equivalently, the condition is that $[M_p]$ belongs to $\text{Outcent}(R_p\Omega)$. This guarantees invertibility of M [RMO, Exercise 38.4]. Indeed, *any bimodule* M *locally free of rank one on the left is invertible:* M is clearly a projective generator on the left, and the map $\Lambda \to \text{Hom}_\Lambda(M,M)$ is an isomorphism locally, hence an isomorphism. Thus

$$\text{Outcent}(\Omega) \subseteq \text{LFP}(\Omega) \subseteq \text{Picent}(\Omega).$$

By (3.3) $\text{LFP}(\Omega)$ *coincides with* $\text{Outcent}(\Omega)$ *if* R *is semilocal.* On the other hand, $\text{LFP}(\Omega)$ coincides with $\text{Picent}(\Omega)$ in the case Ω is commutative [RMO, p.346], giving the usual Picard group of algebraic geometry and commutative algebra.

We can now state the main result we need from the theory of invertible central bimodules.

(3.4) (Fröhlich's localization sequence) *There is an exact sequence:*

$$1 \to \text{Picent}(Z(\Omega)) \xrightarrow{\tau} \text{Picent}(\Omega) \xrightarrow{\tau'} \prod_{p \in \max(R)} \text{Outcent}(R_p\Omega) \to 1.$$

where max(R) *denotes the maximal ideal spectrum of* R, *and* R_p *is the completion of* R *at* $p \in \max(R)$ [RMO, (37.28)]. *The kernel of* τ' *is contained in* $\text{LFP}(\Omega)$, *and there is an induced exact sequence with* "LFP" *replacing* "Picent" *throughout* [RMO, (38.9)].

If R *is semilocal, then* τ' *is an isomorphism* [RS, 1.2.10] *and induces (through the identification of* LFP *with* Outcent) *an isomorphism*

$$\text{Outcent}(\Omega) \cong \prod_{p \in \max(R)} \text{Outcent}(R_p\Omega).$$

We now turn to the group G from Section 2. *Let* R *be semilocal from now on in this section, with no rational prime divisor of* $|G|$ *a unit in* R. The notation H, σ, G_3, σ_3, ... is as in Section 2.

(3.5) *There exists a* **central** *automorphism* γ_5 *of* RG_5, *which induces* σ *under the map* $RG_5 \to RH$.

In fact, the group H will have order $2^6 \cdot 3$, not divisible by 5, so that σ becomes inner upon localization at primes p-dividing 5, and hence can be extended to an inner automorphism of R_pG_5. At any other prime p, σ can be lifted by making use of an obvious decomposition of the localized G_5 group ring,

$$R_pG_5 \simeq I(Q)R_pG_5 \oplus R_pH,$$

where I(Q) is the augmentation ideal of Q. (γ_5 is σ on R_pH and the identity on the other factor.)

Recall that our group G is the pull-back of the diagram (2.8)

$$\begin{array}{ccc} G & \to & G_5 \\ \downarrow & & \downarrow \\ G_3 & \to & H \end{array}.$$

(3.6) Let now Γ be the pull-back of the diagram (2.9)

$$\begin{array}{ccc} \Gamma & \to & ZG_5 \\ \downarrow & & \downarrow \\ ZG_3 & \to & ZH \end{array},$$

where the two group ring maps are induced from the above. Write

$$R\Gamma = R \otimes_Z \Gamma,$$

this is also a pullback in the analogous diagram over R.

The group automorphism σ_3 (cf. Section 2) induces an automorphism on RG_3, which induces σ on RH, and the automorphism γ_5 defined on RG_5 in (3.4) also induces σ on RH. Consequently, we get from the pull-back an automorphism γ on $R\Gamma$.

Thus, we have proved:

(3.7) *There exists an automorphism* γ *of* $R\Gamma$ *which induces* γ_5 *on* RG_5, σ_3 *on* RG_3 *and* σ *on* RH.

In order to construct the desired automorphism of RG, we need some more insight into the structure of group rings, and thus make a detour:

Let Z be any integral domain of characteristic 0 with field of fractions Q. For a normal subgroup N of the finite group G we denote by $I(N) = I_Z(N)$ the augmentation ideal of ZN. Thus

$$I(N)G = I(N)ZG$$

is the kernel of the natural map $ZG \to ZG/N$. We put $\underline{N} = \sum_{x \in N} x$; then

$$e_N = \frac{1}{|N|} \cdot \underline{N}$$

is a central idempotent in QG.

Let now $N_1,...,N_k$ be normal subgroups of G with relatively prime orders $n_1,...,n_k$. (In the later section these orders will be primes, and we will use the primes themselves rather than the consecutive integers $i = 1,2,...,k$ for routine indexing of other quantities; this causes only minor notational changes.) We put $e_i = e_{N_i}$ and

$$\Lambda(G,N_1,...,N_k) = ZG \cdot \left(\sum_{i=1}^{k} (1-e_i) \right).$$

Theorem 3.8. *There is a commutative diagram with exact rows*

$$\begin{array}{ccccccccc} O & \to & \bigcap_{i=1}^{k} I(N_i)G & \to & ZG & \to & \bigoplus_{i=1}^{k} ZG/N_i & & \\ & & \downarrow & & \downarrow & & \downarrow & & \\ O & \to & \bigcap_{i=1}^{k} I(N_i)G & \to & \Lambda & \to & \bigoplus_{i=1}^{k} \hat{\Lambda}_i/n_i\hat{\Lambda}_i & \to & O. \end{array}$$

Here $\Lambda = \Lambda(G,N_1,\ldots,N_k)$, *and*

$$\Lambda_{\hat{i}} = \Lambda(G/N_i,N_1,\ldots,\hat{N}_i,\ldots,N_k), \quad 1 \le i \le k,$$

where, as usual, $\hat{X}$ *means that* X *is omitted. The unlabelled vertical maps are the obvious surjections arising from the definitions of* Λ *and the* Λ_i.

Remark. To assert that a commutative diagram such as the above has exact rows is equivalent to saying that the right-hand square is a pull-back of Z-algebras. In the case k=1, this reduces to the well-known pull-back diagram

$$\begin{array}{ccc} ZG & \rightarrow & ZG/N \\ \downarrow & & \downarrow \\ \Lambda & \rightarrow & Z/nZ(G/N) \end{array},$$

in which all maps are surjective.

We shall next use this result to construct an automorphism of RG. We use $Z = \mathbf{Z}$ for the integral domain above. As subgroups N_i in (3.8) we choose N and Q, and, to make the notation suggestive, we put $N_3 = N$ and $N_5 = Q$. Since there are just two indices, 3 and 5, *we write* Λ_3 *in place of* $\Lambda_{\hat{5}}$, *and* Λ_5 *for* $\Lambda_{\hat{3}}$ in the notation of (3.8). We also write $I(Q) = I_Z(Q)$ for the augmentation ideal of ZQ, and use the subscript (3) for localization at the prime 3, etc.

Then the exact sequences

$$0 \rightarrow I_{(3)}(Q)G_5 \rightarrow \mathbf{Z}_{(3)}G_5 \rightarrow \mathbf{Z}_{(3)}H \rightarrow 0 \qquad (3.9)(a)$$

and

$$0 \rightarrow I_{(5)}(N)G_3 \rightarrow \mathbf{Z}_{(5)}G_3 \rightarrow \mathbf{Z}_{(5)}H \rightarrow 0 \qquad (3.9)(b)$$

are two-sided split.

Hence, with the notation above, we have

$$\Lambda_5/3\cdot\Lambda_5 \cong I_{(3)}(Q)G_5/(3\cdot I_{(3)}(Q)G_5),$$

and (3.10)

$$\Lambda_3/5\cdot\Lambda_3 \cong I_{(5)}(N)G_3/(5\cdot I_{(5)}(N)G_3).$$

Of course *the analogues of* (3.9) *and* (3.10) *for* R *are also valid,* and are obtained from the above $\mathbf{Z}$ versions just by tensoring with R. It is worth observing that

$$R \otimes (\Lambda_5/3\cdot\Lambda_5) \cong (R\Lambda)_5/3\cdot(R\Lambda)_5.$$

We remark that these latter R-algebras are both isomorphic to $\Lambda_5/3\cdot\Lambda_5$ in the basic case where $R = \mathbf{Z}_\pi$ is a semilocalization of $\mathbf{Z}$ (at a set π of primes containing the prime divisors 2, 3, 5 of the group order $|G|$.)

(3.11) *The natural sequence*

$$0 \to I(N)G \cap I(Q)G \to \mathbf{Z}G \to \Gamma \to 0$$

is exact.

With the notation of (3.8), we have a commutative diagram with exact rows

$$\begin{array}{ccccccccc} 0 & \to & I(N)G \cap I(Q)G & \to & \mathbf{Z}G & \to & \Gamma & \to & 0 \\ & & \downarrow \mathrm{id} & & \downarrow & & \downarrow \phi & & \\ 0 & \to & I(N)G \cap I(Q)G & \to & \Lambda & \to & \Lambda_5/3\cdot\Lambda_5 \oplus \Lambda_3/5\cdot\Lambda_3 & \to & 0, \end{array}$$

where ϕ is induced from the two natural compositions

$$\mathbf{Z}G_3 = \mathbf{Z}G/N_5 \to \Lambda_3 \to \Lambda_3/5\cdot\Lambda_3$$

and

$$\mathbf{Z}G_5 = \mathbf{Z}G/N_3 \to \Lambda_5 \to \Lambda_5/3\cdot\Lambda_5.$$

Similar statements hold over R. Moreover, the automorphism γ of $R\Gamma$ constructed in (3.7) above induces σ_3 on $R \otimes (\Lambda_3/5\cdot\Lambda_3)$ and γ_5 on $R \otimes (\Lambda_5/3\cdot\Lambda_5)$, through ϕ and the above maps. Finally, γ_5 is inner as an automorphism of the latter R-algebra.

(3.12) In order to construct an automorphism α_{RG} of RG which induces γ on $R\Gamma$, it is enough to construct an automorphism λ on $R\Lambda$ which induces σ_3 on $R \otimes (\Lambda_3/5\cdot\Lambda_3)$ and γ_5 on $R \otimes (\Lambda_5/3\cdot\Lambda_5)$.

To construct such an automorphism λ, we first consider the automorphism of $\mathbf{Z}G$ induced by the group automorphism σ_G (a lift of σ to G, described in Section 2). Projecting onto Λ we get there an automorphism $\tau : \Lambda \to \Lambda$. Though σ_G is not at all central, a detailed analysis of the inertia groups of characters in Λ, using the explicit structure of G, shows that:

(3.13) *The automorphism τ is central on Λ.*

This is a key property, and was carefully built into the design of G. It gives us a *central* automorphism of $R\Lambda$ which induces $\bar{\sigma}_3$ on $\Lambda_3/5\Lambda_3$. True, it does not give the desired automorphism $\bar{\gamma}_5$ on $\Lambda_5/3\Lambda_5$. The latter automorphism, however, is inner (because 3 does not divide the order of G_5, and γ_5 is central). So we can use the "prime at a time" principal to obtain a central automorphism λ on $R\Lambda$, which agrees with σ_G, up to an inner automorphism, upon localization at 5, and is inner at the other primes. Adjusting λ with the Chinese Remainder Theorem we may assume that it induces $\overline{\sigma}_3$ on $\Lambda_3/5\Lambda_3$ and $\overline{\gamma}_5$ on $\Lambda_5/3\Lambda_5$. Because of (3.12), λ determines an automorphism α_π of RG which induces $\alpha_{R\Gamma}$ on $R\Gamma$. By (2.13)

$$\mathrm{obs}(\alpha_\pi) = \mathrm{obs}(\alpha_{R\Gamma}) = [\sigma] \neq [1].$$

Again by (3.13), the automorphism $\alpha_\pi = \alpha_\pi(\lambda)$ is a counterexample to the semilocal version of the Zassenhaus conjecture:

(3.14) *The automorphism α_π of $\mathbb{Z}_\pi G$ does not have a Zassenhaus factorization. That is, α_π cannot be written as a product $\beta\delta$ of automorphisms with δ central and β induced by a group automorphism. The image group $\alpha_\pi(G)$ is not conjugate to* G *by a unit of* QG.

4. Sketch of the global construction

Having come this far, we now attack the global case. In the semilocal case, our use of invertible bimodules was only implicit, though behind the properties of central automorphisms we used. In the global case we must deal much more directly with these bimodules.

As observed in Bass's Morita theory [B], and reviewed above, every automorphism ν of a ring A gives rise to an invertible bimodule ${}_1A_\nu$ where A acts on the left on itself in the usual way, but on the right through the automorphism ν. Two such bimodules are isomorphic iff the automorphisms defining them differ only by multiplication by an inner automorphism. An invertible bimodule arises from an automorphism iff it is free (of rank 1) on one side (left or right, equivalently). An automorphism is central iff its associated bimodule is central (meaning that the centre of the ring acts in the same way on both sides). A main point of Fröhlich's theory, cf. (3.4), is that invertible central bimodules, for orders in a separable algebra over any Dedekind domain, can be constructed "a prime at a time", using general techniques in integral representation theory. The same methods allow any semilocal automorphism, central or not, to be represented globally by an invertible bimodule. More precisely, in our situation, we have the following:

(4.1) *There is an invertible bimodule* M *for* $\mathbb{Z}G$, *such that* $\mathbb{Z}_\pi \otimes_{\mathbb{Z}} \simeq {}_1(\mathbb{Z}_\pi G)_{\alpha_\pi}$ *as invertible bimodules. There does not exist an automorphism* β *of* G *such that* ${}_1M_\beta$ *is a central bimodule.*

The problem now is that M as left ZG-module might not be free - at least we can not see any obvious reason for this, and it would appear to be a delicate arithmetic issue. If M happens to be free on the left, then M arises from an automorphism which is a counterexample to the Zassenhaus conjecture (1.1) for ZG. The module M is in any event projective on the left, and by (2.8) that G satisfies the Eichler condition [RMO, p.344, (ii)]. By a theorem of Jacobinski [RMO, 38.2] the issue of whether M is free depends only upon the element of the projective class group $K_0(ZG)$ defined by M.

It is perhaps worth pointing out at this point that Jacobinski's work implies there is a ring R of algebraic integers in a suitable algebraic number field K such that $R \otimes_Z M$ is free as left RG-module. Thus we obtain the desired automorphism and counterexample for RG without further work, and this is certainly a global case of some kind.

However, we want to construct a counterexample for the Zassenhaus conjecture for ZG. A large part of the work is devoted to showing that we indeed can find an invertible bimodule M, as above, which is free as left ZG-module and such that

$$Z_\pi \otimes_Z M \cong {}_1(Z_\pi G)_{\alpha_\pi}$$

for some choice of $\alpha_\pi = \alpha_\pi(\lambda)$ as constructed in Section 3. The techniques involve explicitly constructing bimodules, corresponding to the various stages in the construction of α_π, as pull-backs of known bimodules. Using Mayer-Vietoris methods, we then explicitly compute the relevant elements of K_0. In some cases we have to make delicate adjustments to the bimodule, preserving its semilocal structure, to kill these elements of the projective class group. The arguments here have inspired in [GR, Theorem 1.6] a general Mayer-Vietoris sequence for invertible bimodules.

The question remains open as to how generally it is true that a semilocal automorphism of a group ring (or a more general order) is represented by a global one, up to a semilocal inner automorphism. Our arguments here show that at least one choice of $\alpha_\pi = \alpha_\pi(\lambda)$ can be so represented, but perhaps they all can be. Or perhaps there is something special about the collection of automorphisms $\alpha_\pi(\lambda)$. Beyond the desire to obtain a small group easy to work with, the design of our group G was guided entirely by semilocal considerations. In spite of this, and in spite of the many detailed choices required in the final construction, we never later encountered an obstruction in the global situation requiring any revision of the group. Was this due to chance, or is there a general theorem here?

Another important question left open by this paper is the group ring isomorphism problem itself. Indeed, too many good results have been proved now, for the isomorphism problem and the Zassenhaus conjecture, to regard either as settled once and for all by a counterexample.

References

[B] H Bass, *Algebraic K-theory* (Benjamin, New York, 1968).

[B2] H Bass, *AMS Colloquium Lectures* (Winter meeting at Atlanta, January 1978).

[CR] C W Curtis & I Reiner, *Representation theory of finite groups and associative algebras* (John Wiley, New York, 1962).

[CRM] C W Curtis & I Reiner, *Methods of representation theory, I* (John Wiley, New York, 1981).

[Fr] A Fröhlich, The Picard group of noncommutative rings, in particular, of orders, *Trans. Amer. Math. Soc.* **180** (1973), 1-95.

[GR] W Gustafson & K W Roggenkamp, A Mayer-Vietoris sequence for Picard groups, Reiner memorial volume of the *Illinois J.* **32** (1988), 375-406.

[JM] S Jackowski & Z Marchiniak, Group automorphisms inducing the identity map on cohomology, *J. Pure Appl. Algebra* **44** (1987), 241-250.

[Mi] J Milnor, Introduction to algebraic K-theory, *Ann. of Math. Stud.* **72** (1971).

[PS] S Passi & S K Sehgal, Lecture by Passi on torsion units, Manchester conference on representations of finite groups.

[RMO] I Reiner, *Maximal orders* (Academic Press, 1975).

[R] K W Roggenkamp, *Some new progress on the isomorphism problem for integral group rings* MS 1986 (Proceedings of the ring theory conference in Granada 1987, Springer LNM **1328**, 1988), 227-236.

[RS] K W Roggenkamp & L L Scott, Isomorphisms of p-adic group rings, *Ann. of Math.* **126** (1987), 593-647.

[RS1] K W Roggenkamp & L L Scott, On a conjecture of Zassenhaus for finite group rings, manuscript, 1988.

[RS2] K W Roggenkamp & L L Scott, A strong answer to the isomorphism problem for some finite group rings, preliminary manuscript, 1987.

[Sah] C H Sah, Automorphisms of finite groups, *J. Algebra* **10** (1968), 47-68.

[San] R Sandling, *The isomorphism problem for group rings, a survey* (Proceedings of the 1984 Oberwolfach conference on Orders and their applications, Springer LNM **1148**, 1985), 256-289.

[S] L Scott, Recent progress on the isomorphism problem, *Proc. Sympos. Pure Math.* **47** (1987), 259-273.

[Se] S K Sehgal, *Torsion units in integral group rings* (Proc. Nato Institute on Methods in Ring Theory, Antwerp, D. Reidel, Dordrecht, 1983), 497-504.

[W] G E Wall, Finite groups with class-preserving outer automorphisms, *J. London Math. Soc.* **22** (1947), 315-320.

[War] H N Ward, *Some results on the group algebra over a prime field* (Mimeo. notes, Harvard University, 1960-61).

[Z] H Zassenhaus, On the torsion units of group rings, *Studies in Math., Inst. de alta Cultura, Lisboa* (1974), 119-126.

[Z2] Personal copy of L Scott, of a preprint of [Se], with conjectures attributed to Hans Zassenhaus personally initialled by Zassenhaus.

THE FIBONACCI GROUPS REVISITED

RICHARD M THOMAS

University of Leicester, Leicester LE1 7RH

Dedicated to the memory of R C Lyndon

1. Introduction

The purpose of this paper is to survey some results concerning our old friends, the Fibonacci groups; as with many surveys, the choice of results referred to is probably rather subjective. To begin with, recall that the Fibonacci group F(2,n) is the group defined by the presentation

$$< a_1, a_2, \ldots\ldots, a_n : a_1a_2 = a_3, a_2a_3 = a_4, \ldots\ldots, a_{n-1}a_n = a_1, a_na_1 = a_2 >.$$

The study of these groups began in earnest after a question of Conway [Con65] as to whether or not F(2,5) is cyclic of order 11, and it was quickly determined in [Con67] that this was indeed the case, and also that F(2,1) and F(2,2) are trivial, F(2,3) is the quaternion group of order 8, F(2,4) is cyclic of order 5, and F(2,6) is infinite. F(2,7) was shown to be cyclic of order 29 using a computer, as was reported in [Bru74] and [ChJ77], and an algebraic proof, based on a coset enumeration, given in [Hav76]; the latter is certainly non-trivial, and, whilst it is pointed out in [Lee84] that "there exists a formal proof half as long as Havas's proof", it does seem that this class of groups provides, at the very least, some good tests for implementations of the Todd-Coxeter algorithm and the ingenuity of group theorists! An alternative approach, which uses the term-rewriting system REVE to find the order of F(2,5) and to prove that F(2,6) is infinite, is described in [Mar86].

However, F(2,8) and F(2,10) were shown to be infinite in [Bru74], and F(2,9) was shown to have order at least 152.5^{741} in [HRS79], and recently to be infinite also, in [New89]. A major contribution is an unpublished result of Lyndon, who used small cancellation theory to show that F(2,n) is infinite for $n \geq 11$; we should like to dedicate the present paper to the memory of Roger Lyndon.

We can generalize the above by defining F(r,n) to be the group defined by the presentation

$$< a_1, a_2, \ldots\ldots , a_n \; : \; a_1a_2\ldots a_r = a_{r+1},\; a_2a_3\ldots a_{r+1} = a_{r+2}, \ldots , \; a_{n-1}a_na_1\ldots a_{r-2} = a_{r-1},\; a_na_1a_2\ldots a_{r-1} = a_r >,$$

where r > 0, n > 0, and all subscripts are assumed to be reduced modulo n. If r = 1, we have the infinite cyclic group and, if n = 1 with r > 1, we have a cyclic group of order r - 1; so we shall assume throughout that r > 1 and that n > 1. It was shown in [Joh74a] that the derived quotient F(r,n)/F(r,n)' is always finite, in [JWW74] that F(r,n) is cyclic of order r - 1 if n divides r, and in [CaR74a] and [CaR75b] that

(1.1) F(r,n) is metacyclic of order $r^n - 1$ if $r \equiv 1 \pmod n$.

On the other hand, it was shown in [Tho83] and [CaT86] that

(1.2) F(r,n) is infinite if (r+1, n) > 3, or if (r+1, n) = 3 with n even or r > 2,

where (a, b) denotes the greatest common divisor of a and b. In particular, we have that

(1.3) F(r,2) is cyclic of order r - 1 if r is even, and metacyclic of order $r^2 - 1$ if r is odd,

which had already been noted in [JWW74], and that

(1.4) F(r,3) is cyclic of order r - 1 if $r \equiv 0 \pmod 3$, metacyclic of order $r^3 - 1$ if $r \equiv 1 \pmod 3$, and infinite if $r \equiv 2 \pmod 3$ with r > 2.

An alternative proof of the fact that F(3k-1,3) is infinite for k > 1 may be found in [Sea82]. The order of F(4k+2,4) was also determined in [Sea82], where the groups were shown to be metabelian, and these groups were shown to be metacyclic in [Tho89b]; for further details, see Section 3. So we have that

(1.5) F(r, 4) is cyclic of order r - 1 if $r \equiv 0 \pmod 4$, metacyclic of order $r^4 - 1$ if $r \equiv 1 \pmod 4$, metacyclic of order $(4k + 1)\,[2^{4k+1} + (-1)^k 2^{2k+1} + 1]$ if $r \equiv 2 \pmod 4$, and infinite if $r \equiv 3 \pmod 4$.

We do not, however, have a similar result covering the groups F(r,5); for example, it does not seem to be known whether or not the group F(7,5) is finite. In addition to the results already mentioned, the use of small cancellation theory in [ChJ77] gave that

(1.6) if n does not divide any of $r \pm 1$, $r \pm 2$, $2r \pm 1$, 3r, 4r or 5r, then F(r,n) is infinite, and this was generalized in [Sea82] to give that

(1.7) if r is odd, and if n does not divide any of $r \pm 1$, $r + 2$, $2r$, $2r + 1$ or $3r$, then F(r,n) is infinite;

(1.8) if $s \geq 0$ such that 2^s divides (r, n), 2^{s+1} does not divide r, and n does not divide any of $r \pm 1$, $r + 2$, $2r$, $2r + 1$ or $3r$, then F(r,n) is infinite;

and

(1.9) if n does not divide any of $r \pm 2$, $r \pm 3$, $2r$, $2r \pm 1$, $2r \pm 2$ or $3r \pm 1$, then F(r,n) is infinite.

As a consequence, we have that

(1.10) if $n > 2r + 1$, then F(r,n) is infinite unless $r = 2$, $n = 7$ or (possibly) $r = 3$, $n = 9$.

The situation with regards to our knowledge as to which of the groups F(r,n) with $2 \leq r \leq 10$ and $2 \leq n \leq 10$ are finite is summarised below :

n :	2	3	4	5	6	7	8	9	10
r = 2	1	8	5	11	∞	29	∞	∞	∞
3	8	2	∞	22	1512	?	∞	?	∞
4	3	63	3	∞	?	?	?	?	∞
5	24	∞	624	4	∞	?	∞	∞	?
6	5	5	125	7775	5	∞	?	?	∞
7	48	342	∞	?	$7^6 - 1$	6	∞	?	∞
8	7	∞	7	?	∞	$8^7 - 1$	7	∞	?
9	80	8	6560	∞	?	∞	$9^8 - 1$	8	∞
10	9	999	4905	9	?	?	∞	$10^9 - 1$	9

Apart from the fact that F(2,9) is infinite, the situation seems to be the same as was reported in [Sea82]. All the cases where the groups are known to be infinite in the table follow from the above results; the cases where the groups are finite may be checked by the above results and coset enumeration. The only finite entry in the above table that is not metacyclic is the group F(3,6), whose structure was determined in [CaR74b]. In general, for a nice survey of results concerning these groups, see [ChJ77], [Joh76], [Joh80] and [JWW74]; also, a survey of these and some related groups may be found in [CRT87].

2. Representations

Another proof of the result of Brunner and Lyndon that the groups F(2,2m) are infinite for $m \geq 4$ is given in [Tho89a], where a homomorphism from F(2,2m) onto an infinite subgroup of PSL(2,C) is exhibited. In this section, we show how this representation can be deduced in a natural way, and determine the connection between this approach and the recent work of Helling, Kim and Mennicke in [HKM89]. In general, the group F(r,n) admits an automorphism of order n which permutes the generators cyclically, and, if we form a semi-direct product of F(r,n) with a cyclic group of order n acting in this way, we get the group E(r,n) with presentation

$$< x, t : xt^r = tx^r, t^n = 1 >,$$

where x denotes ta_1^{-1}; see [Joh74b] or [JWW74] for example. Let E := E(2,2m), a:= x^2 and b:= t^{-2}; then it is shown in [Tho89a] that N = < a, b > is a normal subgroup of index 2 in E with presentation

$$< a, b, c : a^m = b^m = 1, c = a^{-1}b^{-1}ab, cac^{-1} = b^{-1} >.$$

Now N admits an automorphism of order 2 which interchanges a and b and inverts c, and we can form the semi-direct product of N with a cyclic group < s > of order 2 to get the group L with presentation

$$< a, s, c : a^m = s^2 = 1, c = [a, s^{-1}as], cac^{-1} = s^{-1}a^{-1}s >.$$

We introduce v := sc, and then delete $c = s^{-1}v = sv$, to get

$$< a, s, v : a^m = s^2 = 1, sv = [a, s^{-1}as], vav^{-1} = a^{-1} >.$$

The relation sv = [a, sas] is equivalent to $v = s[a, sas] = (as)^{-2}s(as)^2$, and we may use this to delete the generator v and get

$$< a, s : a^m = s^2 = ((a^{-1}s)^2(as)^3)^2 = 1 >.$$

We now want to find matrices A and S in PSL(2,C) such that

$$A^m = S^2 = ((A^{-1}S)^2(AS)^3)^2 = I.$$

Let $S = \begin{pmatrix} \lambda & \mu \\ \nu & -\lambda \end{pmatrix}$, where $\lambda^2 + \mu\nu = -1$, $\alpha = e^{\pi i/m}$ and $A = \begin{pmatrix} \alpha & 0 \\ \beta & \alpha^{-1} \end{pmatrix}$ for some, as yet unspecified, value of β. We have $A^m = S^2 = -I$ in SL(2,C), and we would like to choose λ, μ, ν and β such that $(A^{-1}S)^2(AS)^3$ has trace 0, for then we would have that $((A^{-1}S)^2(AS)^3)^2 = -I$ in SL(2,C). Now, in general, we have that

$$\mathrm{Tr}\,(UV) + \mathrm{Tr}\,(U^{-1}V) = \mathrm{Tr}\,(U)\,\mathrm{Tr}\,(V) \qquad (2.1)$$

for any matrices U and V in SL(2,C), and we may use this to deduce that

$$\mathrm{Tr}\,(U^2V^3) = [\mathrm{Tr}\,(UV)\,\mathrm{Tr}\,(U) - \mathrm{Tr}\,(V)]\,[\mathrm{Tr}\,(V)^2 - 1] - [\mathrm{Tr}\,(U)^2 - 2]\,\mathrm{Tr}\,(V). \qquad (2.2)$$

Setting $U = A^{-1}S$, $V = AS$ and $\phi = \mathrm{Tr}\,(V)$, and then noting that

$$\mathrm{Tr}\,(U) = \mathrm{Tr}\,(SU^{-1}S^{-1}) = \mathrm{Tr}\,(VS^{-2}) = -\phi,$$

we have that

$$\mathrm{Tr}\,((A^{-1}S)^2(AS)^3) = -\phi\,[\mathrm{Tr}\,(A^{-1}SAS) + 1]\,[\phi^2 - 1] - \phi\,[\phi^2 - 2] \qquad (2.3)$$

from (2.2), and so, assuming that $\phi \neq 0$, we want that $\mathrm{Tr}\,(A^{-1}SAS) = \dfrac{3 - 2\phi^2}{\phi^2 - 1}$.

Now (2.1) gives that $\mathrm{Tr}\,(A^{-1}SAS) = \mathrm{Tr}\,(A^{-1}S)\,\mathrm{Tr}\,(AS) - \mathrm{Tr}\,(A^2)$, and so we have

$$\frac{3 - 2\phi^2}{\phi^2 - 1} = -\phi^2 - \alpha^2 - \alpha^{-2}. \qquad (2.4)$$

Solving (2.4) gives that ϕ must satisfy $\phi^4 + \zeta\phi^2 - \zeta = 0$, where $\zeta = 2\cos\left(\frac{2\pi}{m}\right) - 3$. Let

$$\rho = -\zeta^2 - 4\zeta = \left(3 - 2\cos\left(\frac{2\pi}{m}\right)\right)\left(1 + 2\cos\left(\frac{2\pi}{m}\right)\right),$$

so we may take ϕ^2 to be $\dfrac{-\zeta + i\sqrt{\rho}}{2}$. We get a map θ from L into PSL(2,C) by taking $a\theta = A := \begin{pmatrix} \alpha & 0 \\ \phi & \alpha^{-1} \end{pmatrix}$, $s\theta = S := \begin{pmatrix} 0 & 1 \\ -1 & 0 \end{pmatrix}$. Let $B := S^{-1}AS = \begin{pmatrix} \alpha^{-1} & -\phi \\ 0 & \alpha \end{pmatrix}$ so that $N\theta = \langle A, B \rangle$ is a representation of N in PSL(2,C). We now get a representation $\langle A_1, A_3 \rangle$ of

$$F = \langle a_1, a_2 \rangle = \langle a_1, a_1a_2 \rangle = \langle a_1, a_3 \rangle$$

by noting that

$$a_1 = x^{-1}t = t^2t^{-2}x^{-1}t = t^2x^{-2}t^{-1}t = b^{-1}a^{-1}$$

and

$$a_3 = t^{-2}a_1t^2 = t^{-2}x^{-1}t^3 = x^{-2}t^{-1}t^3 = a^{-1}b^{-1},$$

and then taking

$$A_1 = \begin{pmatrix} 1 - \phi^2 & \alpha\phi \\ -\alpha^{-1}\phi & 1 \end{pmatrix}, \quad A_3 = \begin{pmatrix} 1 & \alpha^{-1}\phi \\ -\alpha\phi & 1 - \phi^2 \end{pmatrix}.$$

If $m \geq 4$, then $\rho > 0$, and A_1 and A_3 both have

$$\text{trace}\,\gamma := 2 - \phi^2 = \left(\frac{\cos(2\pi/m) + 1 - i\sqrt{\rho}}{2}\right);$$

we may calculate directly that $A_2 := A_1^{-1}A_3$ and $A_4 := A_2A_3$ both have

$$\text{trace}\,\delta := \phi^2 + \zeta + 2 = \left(\frac{\cos(2\pi/m) + 1 + i\sqrt{\rho}}{2}\right),$$

and, since ϕ^2 is not real, none of these matrices has finite order; in particular, F is infinite. On the other hand, if m = 2, then F is cyclic of order 5, and the representation is faithful; if m = 3, then Fθ has order 4, whilst F is infinite; see [Tho88] for details.

Alternatively, it was shown in [HKM89] (see also [Men89]) that there is a tessellation of the 2-sphere consisting of 4m triangles, 6m edges and 2m + 2 vertices with each oriented edge labelled by one of $\{a_1, a_2, \ldots, a_{2m}\}$, each label ocurring precisely three times, and where each triangle has a boundary labelled $x_j, x_{j+1}, x_{j+2}^{-1}$ for some j (mod 2n); for example, if n = 5, we get a labelled icosahedron as shown on the right.

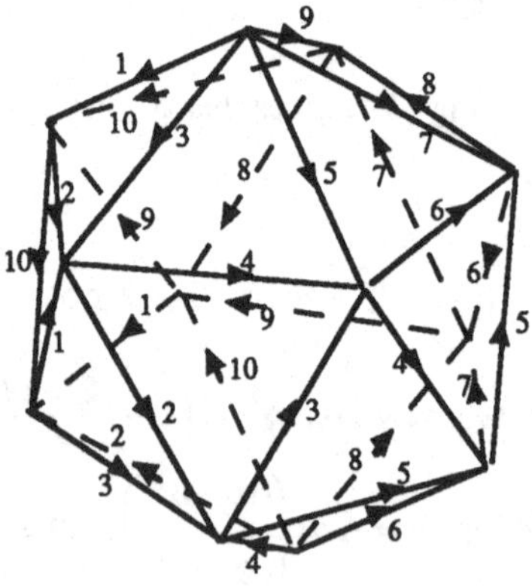

Given this, F(2,2m) acts as a group of isometries of a semi-regular tessellation of hyperbolic 3-space whose fundamental domain is a semiregular polyhedron which is a metric realization of the combinatorial polyhedron described above. This shows, among other things, that F(2,2m) is represented faithfully as a discrete subgroup of SL(2,C). It is shown in [HKM89] that any representation ψ of F(2,2m) in SL(2,C) which satisfies

$$\mathrm{Tr}\,(a_1\psi) = \mathrm{Tr}\,(a_3\psi) = \gamma \text{ and } \mathrm{Tr}\,(a_2\psi) = \mathrm{Tr}\,(a_4\psi) = \delta$$

is determined uniquely up to conjugacy within SL(2,C). This shows that the representation described above is equivalent to the one described in [HKM89], and, in particular, that it is faithful.

3. Generalizations

The have been various generalizations of the Fibonacci groups, for example the groups H(r,n,s) defined by the presentations

$$< a_1, a_2, \ldots\ldots , a_n : a_1a_2\ldots a_r = a_{r+1}a_{r+2}\ldots a_{r+s}, a_2a_3\ldots a_{r+1} = a_{r+2}a_{r+3}\ldots a_{r+s+1}, \ldots\ldots , a_{n-1}a_na_1\ldots a_{r-2} = a_{r-1}a_r\ldots a_{r+s-2}, a_na_1a_2\ldots a_{r-1} = a_ra_{r+1}\ldots a_{r+s-1} >,$$

where $r > s \geq 1$, and the groups F(r,n,k) defined by the presentations

$$< a_1, a_2, \ldots\ldots , a_n : a_1a_2\ldots a_r = a_{r+k}, a_2a_3\ldots a_{r+1} = a_{r+k+1}, \ldots\ldots , a_{n-1}a_na_1\ldots a_{r-2} = a_{r+k-2}, a_na_1a_2\ldots a_{r-1} = a_{r+k-1} >,$$

where $r \geq 2$ and $k \geq 0$; as usual, all subscripts are taken to be reduced modulo n. Clearly, the groups H(r,n,1) and F(r,n,1) are each isomorphic to F(r,n). The groups H(r,n,s) were introduced in [CaR75a], and it was proved there that H(r,n,s) is metacyclic of order $r^n - s^n$ if $r \equiv s \pmod n$ and (r, n) = (r, s) = 1, which is a generalization of (1.1), and conjectured that H(r,4,2) is metacyclic for r odd; this was proved to be the case in [Bru77].

As mentioned in Section 1, the groups F(4k+2,4) were proved in [Sea82] to be metabelian of order

$$(4k + 1)(2^{4k+1} + (-1)^k 2^{2k+1} + 1),$$

and this is the same as the order of the group H(4k+3,4,2); it was conjectured in [Sea82] that F(4k+2,4) is, in fact, isomorphic to H(4k+3,4,2), and hence metacyclic, and this was proved to be the case in [Tho89b]. On the other hand, it was shown in [CaT86] that

(3.1) H(r,n,s) is infinite if (r+s, n) > 3, or if (r+s, n) = 3 with r+s > 3, or if (r+s, n) > 1 with (r, s) > 1,

which generalizes [Tho83] and some of the results in [CaR75a]. With regard to the groups F(r,n,k), it was shown in [Joh75] that

(3.2) F(r,n,k) is infinite if (r-1, n, k) > 1 or if $\nu_2(r+1) > \nu_2(k-1) < \nu_2(n)$, where $\nu_2(m) := \alpha$ if $m = 2^\alpha q$ with q odd and $\nu_2(0) := \infty$, and in [CaT86] that

(3.3) if d = (n, k-1) and F(r,d) is infinite, then F(r,n,k) is infinite,

which, combined with (1.2), gives that

(3.4) if (r+1, n, k-1) > 3, or if (r+1, n, k-1) = 3 with n even or n > 2, then F(r,n,k) is infinite.

This generalizes the result in [CaR74b] that the groups F(2λt-1, μt, 1-νt) are infinite for $t \geq 4$. On the other hand, it was shown in [CaR75b] that, if $r \equiv 1 \pmod n$ and (k, n) = 1, then F(r,n,k) is isomorphic to F(r,n), and hence is metacyclic of order $r^n - 1$, and in [CaR74b] that F(2,5,2) is isomorphic to SL(2,5); another proof of this latter fact appears in [JoM75]. Further generalizations may be found in [CaR75c], where another such non-metacyclic group of order 1512, not isomorphic to F(3,6) is described, and [CaR79].

Acknowledgements

The author would like to pay tribute to all those who have contributed to our understanding of the Fibonacci groups, and, in particular, to Colin Campbell, Dave Johnson and Edmund Robertson, both for their work in this area and also for their friendship and encouragement over several years; thanks are also due to Hilary Craig for all her encouragement and forebearance and to Fidel Baptista for his technical assistance.

References

[Bru74] A M Brunner, The determination of Fibonacci groups, *Bull. Austral. Math. Soc.* **11** (1974), 11-14.

[Bru77] A M Brunner, On groups of Fibonacci type, *Proc. Edinburgh Math. Soc.* **20** (1977), 211-213.

[CaR74a] C M Campbell & E F Robertson, The orders of certain metacyclic groups, *Bull. London Math. Soc.* **6** (1974), 312-314.

[CaR74b] C M Campbell & E F Robertson, Applications of the Todd-Coxeter algorithm to generalized Fibonacci groups, *Proc. Royal Soc. Edinburgh* **73A** (1974/5), 163-166.

[CaR75a] C M Campbell & E F Robertson, On a class of finitely presented groups of Fibonacci type, *J. London Math. Soc.* **11** (1975), 249-255.

[CaR75b] C M Campbell & E F Robertson, On metacyclic Fibonacci groups, *Proc. Edinburgh Math. Soc.* **19** (1975), 253-256.

[CaR75c] C M Campbell & E F Robertson, A note on Fibonacci type groups, *Canad. Math. Bull.* **18** (1975), 173-175.

[CaR79] C M Campbell & E F Robertson, Finitely presented groups of Fibonacci type. Part II, *J. Austral. Math. Soc.* **28A** (1979), 250-256.

[CRT87] C M Campbell, E F Robertson & R M Thomas, Fibonacci numbers and groups, in *Applications of Fibonacci Numbers* (eds. A F Horadam, A N Philippou & G E Bergum, D Reidel, 1987), 45-59.

[CaT86] C M Campbell & R M Thomas, On infinite groups of Fibonacci type, *Proc. Edinburgh Math. Soc.* **29** (1986), 225-232.

[ChJ77] C P Chalk & D L Johnson, The Fibonacci groups II, *Proc. Royal Soc. Edinburgh* **77A** (1977), 79-86.

[Con65] J H Conway, Advanced problem 5327, *Amer. Math. Monthly* **72** (1965), 915.

[Con67] J H Conway et al., Solution to advanced problem 5327, *Amer. Math. Monthly* **74** (1967), 91-93.

[Hav76] G Havas, Computer aided determination of a Fibonacci group, *Bull. Austral. Math. Soc.* **15** (1976), 297-305.

[HRS79] G Havas, J S Richardson & L S Sterling, The last of the Fibonacci groups, *Proc. Royal Soc. Edinburgh* **83A** (1979), 199-203.

[HKM89] H Helling, A C Kim & J L Mennicke, On Fibonacci groups, preprint.

[Joh74a] D L Johnson, A note on the Fibonacci groups, *Israel J. Math.* **17** (1974), 277-282.

[Joh74b] D L Johnson, Extensions of Fibonacci groups, *Bull. London. Math. Soc.* **7** (1974), 101-104.

[Joh75] D L Johnson, Some infinite Fibonacci groups, *Proc. Edinburgh Math. Soc.* **19** (1975), 311-314.

[Joh76] D L Johnson, *Presentation of groups* (London Math. Soc. Lecture Note Series **22**, Cambridge University Press, 1976).

[Joh80] D L Johnson, *Topics in the theory of group presentations* (London Math. Soc. Lecture Note Series **42**, Cambridge University Press, 1980).

[JoM75] D L Johnson & H Mawdesley, Some groups of Fibonacci type, *J. Austral. Math. Soc.* **20A** (1975), 199-204.

[JWW74] D L Johnson, J W Wamsley & D Wright, The Fibonacci groups, *Proc. London Math. Soc.* **29** (1974), 577-592.

[Lee84] J Leech, Coset enumeration, in *Computational group theory* (ed. M D Atkinson , Academic Press, 1984).

[Mar86] U Martin, *Doing algebra with* REVE (Technical Report **UMCS-86-10-4**, Department of Computer Science, University of Manchester, 1986).

[Men89] J L Mennicke, On Fibonacci groups and some other groups, in *Groups- Korea 1988* (ed. A C Kim & B H Neumann, Lecture Notes in Mathematics **1398**, Springer-Verlag, 1989), 117-123.

[New89] M F Newman, *Proving a group infinite* (Research Report **26**, Department of Mathematics, Australian National University, 1988).

[Sea82] D J Seal, The orders of the Fibonacci groups, *Proc. Roy. Soc. Edinburgh* **92A** (1982), 181-192.

[Tho83] R M Thomas, Some infinite Fibonacci groups, *Bull. London Math. Soc.* **15** (1983), 384-386.

[Tho88] R M Thomas, ***Representations of certain subgroups of extended Fibonacci groups*** (Technical Report **15**, Department of Computing Studies, University of Leicester, December 1988).

[Tho89a] R M Thomas, The Fibonacci groups F(2,2m), *Bull. London Math. Soc.* **21** (1989), 463-465.

[Tho89b] R M Thomas, The Fibonacci groups F(4k+2,4), *Comm. Algebra*, to appear.

GALOIS GROUPS

JOHN G THOMPSON

University of Cambridge, Cambridge CB2 1SB

The primary purpose of this note is to present one of the exact sequences of Galois theory. I also place on record a finiteness theorem which is not well known, but which in certain circumstances may be of use. As regards the exact sequence, I have benefitted from conversations with M Fried, and as regards the finiteness theorem, I have benefitted from conversations with D and G Chudnovsky, who have produced rather convincing computational evidence that a conjecture of mine [12] is false.

If G is a group, $\hat{G}$ denotes the profinite completion of G [3, p.156]. If k is a field, G_k denotes the Galois group of $\bar{k}/k$, where $\bar{k}$ is a separable algebraic closure of k. Set $\Gamma = PSL_2(\mathbf{Z})$. The exact sequence in question is the following:

$$1 \to \hat{\Gamma} \to \mathfrak{G} \to G_{\mathbf{Q}} \to 1. \qquad (1)$$

I would like to explain in elementary terms the construction of (1) and its Galois-theoretic interpretation.

Let

$$h = \{x + iy \mid x,y \in \mathbf{R},\ y > 0\},$$

$$\hat{h} = h \cup \mathbf{Q} \cup \{\infty\}.$$

The game we play in a small corner of the gaga playground (gaga is an acronym for géométrie algébrique géométrie analytique) concerns maps from $\hat{h}$ to the extended complex plane $\hat{\mathbf{C}} = \mathbf{C} \cup \{\infty\}$, and involves rudimentary topology and function theory. The topology on $\hat{h}$ is the one for which a basis for open sets is given by

(a) open disks contained in h,

(b) sets $\{q\} \cup D_{q,r}$, where $q \in \mathbf{Q}$, and $D_{q,r}$ is the open disk of radius $r > 0$, centered at $q + ir$,

(c) sets $\{\infty\} \cup \{z = x + iy \in h \mid y > c\}$, where $c > 0$.

One set of functions to be considered is the set $\mathfrak{F}$ of all $\hat{\mathbf{C}}$-valued functions f on $\hat{h}$ which are meromorphic on $\hat{h}$, by which is meant that for each $\tau_0 \in h$, f can be represented by a convergent Laurent series in $\tau - \tau_0$ on some neighbourhood of τ_0, while for each $q = \frac{a}{c} \in \mathbf{Q}$, there is a neighbourhood of q on which f can be represented by a convergent Laurent series in $e_{-Nc^2,q}$ for some positive integer N, and where

$$e_{w,r}(\tau) = \exp\frac{2\pi i}{w(\tau - r)} \quad (w < 0,\ w,\ r \in \mathbf{R}).$$

As for ∞, we demand that on some neighbourhood of ∞, f is represented by a convergent Laurent series in e_N for some $N \in \mathbf{N}$, where $e_N(\tau) = \exp\frac{2\pi i\tau}{N}$. This description of $\mathcal{F}$ is adequate for the present purposes, and one may compare this description with the discussion given, for example, in Schoeneberg's book [7].

Since Γ acts on $\hat{h}$ via

$$\tau \mapsto \frac{a\tau + b}{c\tau + d} = \begin{pmatrix} a & b \\ c & d \end{pmatrix}\tau,$$

where $\begin{pmatrix} a & b \\ c & d \end{pmatrix}^{\cdot} \in \Gamma$, the dot indicating the image in Γ of an element of $SL_2(\mathbf{Z})$, a moment's thought shows that we get an action $\mathcal{F} \times \Gamma \to \mathcal{F}$ of Γ on $\mathcal{F}$ by defining $(f,A) \mapsto f \cdot A$, where $(f \cdot A)(\tau) = f(A\tau)$. The underlying group-theoretic reason for the existence of this action stems from the relation

$$\begin{pmatrix} a & b \\ c & d \end{pmatrix}\begin{pmatrix} 1 & 1 \\ 0 & 1 \end{pmatrix}\begin{pmatrix} d & -b \\ -c & a \end{pmatrix} = \begin{pmatrix} 1-ac & a^2 \\ -c^2 & 1+ac \end{pmatrix}$$

if $\begin{pmatrix} a & b \\ c & d \end{pmatrix} \in SL_2(\mathbf{Z})$, which guarantees that if μ is the set of all roots of unity, then

$$\{\mu e_N \mid N \in \mathbf{N}\} \text{ is carried to } \{\mu e_{-Nc^2,a/c} \mid N \in \mathbf{N}\}$$

by $A = \begin{pmatrix} d & -b \\ -c & a \end{pmatrix}^{\cdot}$. More precisely,

$$\exp\frac{2\pi i}{N}\left(\frac{d\tau - b}{-c\tau + a}\right) = \xi \cdot \exp\frac{2\pi i}{-Nc^2(\tau - \frac{a}{c})},$$

where $\xi = \exp\frac{-2\pi i d}{cN}$ is independent of τ.

The set $\mathcal{F}$ is (probably) too big, so we set

$$\mathcal{F}_0 = \{f \in \mathcal{F} \mid f \circ \Gamma \text{ is finite}\}.$$

$\mathfrak{F}_0$ is a field under pointwise addition and multiplication of functions. Moreover, if Γ_0 is any subgroup of Γ of finite index, it is straightforward to construct f in $\mathfrak{F}_0$ whose stabilizer in Γ is Γ_0. The set of fixed points of Γ on $\mathfrak{F}_0$ is C(j), the field generated by C and j, the elliptic modular function [7, p.36; 8, p.89]. Since $\mathfrak{F}_0$ is a field and $\hat{\Gamma}$ is a group of automorphisms of $\mathfrak{F}_0$ and C(j) is the fixed field of $\hat{\Gamma}$, we see that

$$\hat{\Gamma} \cong \text{Gal}\, \mathfrak{F}_0/\mathbf{C}(j).$$

This is the first step in the construction of (1).

The next step seems initially nonsensical, but it is not. Suppose $f \in \mathfrak{F}_0$. Then there is an irreducible monic $F_f \in \mathbf{C}(T)[U]$ such that $F_f(j,f) = 0$. There is an action Aut $\mathbf{C} \times \mathbf{C}[T,U] \to \mathbf{C}[T,U]$, $(\sigma, \Sigma\, a_{ij}T^iU^j) \mapsto \Sigma\, \sigma(a_{ij})T^iU^j$, which induces an action on $\mathbf{C}(T)[U]$. The trick is to notice that $\sigma(F_f)(j,U)$ has a root in $\mathfrak{F}_0$. I know by direct interrogation of colleagues that this fact is known to some and is opaque to others. A careful, perhaps unduly careful, proof appears in [13].

Let K be the set of elements of $\mathfrak{F}_0$ which are algebraically dependent on Q(j). Then K is a Galois extension of Q(j), and if we set $\mathfrak{G}$ = Gal K/Q(j), then $\hat{\Gamma}$ is isomorphic to the subgroup of $\mathfrak{G}$ whose fixed field is $\bar{\mathbf{Q}}(j)$, and (1) follows immediately. There is a section $G_Q \to \mathfrak{G}$ obtained by letting G_Q act on the "fibre at ∞"; if $f \in K$ and in some neighbourhood of ∞, $f = \Sigma_{n \geq M}\, a_n(f)e_N^n$, then σ in G_Q sends f to that element of K whose behaviour at ∞ is given by $\Sigma\, \sigma(a_n(f))e_N^n$.

Thus, (1) is a split extension.

It is a curious fact that from (1), one quickly shows that the Fischer-Griess group F_1 is a Galois group over Q [13]. Just why this should be so in such a trivial way is a mystery to me.

In a precise sense, now to be explained, $\mathfrak{G}$ is adequate to carry out Galois-theoretic constructions over Q(T). Working in some algebraic closure of Q(T), we say that an extension field E of Q(T) is regular if and only if $E \cap \bar{\mathbf{Q}} = \mathbf{Q}$. Following Serre [9], set

Gal_T = {G | G is a finite group and for some regular Galois extension E of Q(T), $G \cong \text{Gal}\, E/\mathbf{Q}(T)$.}.

It is a consequence of Hilbert's Irreducibility Theorem [5] that if $G \in \text{Gal}_T$, then G is a Galois group over Q. This is a remarkable link, now taken in stride by the young and not-so-young. Another remarkable fact is Belyi's theorem [2], which implies that if $G \cong \mathrm{Gal}\, E/\mathbf{Q}(T)$, E being regular and Galois, then there are fields F_0, F, and an element f of K, such that

$$\mathbf{Q}(j) \subseteq F_0 = \mathbf{Q}(f) \subseteq F \subseteq K,$$

and an isomorphism of E onto F which carries T to f. This is the precise sense in which K is adequate to carry out Galois-theoretic constructions over $\mathbf{Q}(T)$. R Guralnick and I [4] have begun the systematic study of the functions f in K such that $j \in \mathbf{Q}(f)$, but the work is still in its early stages. My idea is first to study those f such that $j \in \bar{\mathbf{Q}}(f)$, and then try to descend to $\mathbf{Q}(f)$. I can't say that this approach has won any fervent adherents apart from me, but there is no particular hurry, for if the idea is a good one, it will carry the day; if it is not, it won't. The relation $j \in \bar{\mathbf{Q}}(f)$ gives rise to a triple (α,β,γ) of elements of S_n, where $n = [\bar{\mathbf{Q}}(f) : \bar{\mathbf{Q}}(j)]$, such that $\alpha^2 = \beta^3 = \alpha\beta\gamma = 1$ and such that $\langle\alpha,\beta,\gamma\rangle$ is transitive. In addition, $c(\alpha) + c(\beta) + c(\gamma) = n+2$, where, for $\sigma \in S_n$, $c(\sigma)$ is the number of orbits of $\langle\sigma\rangle$ on $\{1,\ldots,n\}$. The classification of the orbits of S_n on the set of such triples is hardly feasible, but the additional constraint $j \in \mathbf{Q}(f)$ introduces number-theoretic considerations which should be examined more closely.

The Fuchsian group Γ leads naturally to the set $\mathcal{F}(g;r;e_1,\ldots,e_m)$ of Fuchsian groups $G \subseteq PSL_2(\mathbf{R})$ such that $G\backslash h$ has finite hyperbolic area, genus g, r cusps, and where $e_1,\ldots,e_m$ are the orders of a set of representatives for the conjugacy classes of maximal elliptic subgroups of G, with the conventions that $e_1 \leq \ldots \leq e_m$, and that when G is torsion free, we write $\mathcal{F}(g;r;\phi)$. With this notation, $\Gamma \in \mathcal{F}(0;1;2,3)$. Set $\mathcal{F}(r) = \mathcal{F}(0;r;\phi)$.

If $G \in \mathcal{F}(r)$ and $r \geq 3$, then $G = \langle g_1,\ldots,g_r \mid g_1 \cdot\ldots\cdot g_r = 1\rangle$, $g_i = \{A_i,-A_i\}$, $\mathrm{tr}\, A_i = 2$, and G is free of rank r-1. In addition, the A_i are pairwise conjugate in $SL_2(\mathbf{R})$ [1 and 6]. Let q_i be the unique fixed point of g_i on $\mathbf{R} \cup \{\infty\}$, and set

$$\hat{h} = h \cup \bigcup_{i=1}^{r} G(q_i).$$

For each non negative even integer k, set

$$A_k(G) = \{f \mid f \text{ is holomorphic on } \hat{h} \text{ and } f\left(\frac{a\tau+b}{c\tau+d}\right) = (c\tau + d)^k f(\tau),$$
$$\tau \in \hat{h},\ \begin{pmatrix} a & b \\ c & d \end{pmatrix} \in G\}.$$

Let $A(G)$ be the graded ring whose k^{th} piece is $A_k(G)$. Assume also that $A_1 = \begin{pmatrix} 1 & 1 \\ 0 & 1 \end{pmatrix}$. Then to each element f of $A(G)$ is associated its Fourier series

$$f = \sum a_n(f)e_1^n,$$

convergent in some neighbourhood of ∞. Each $A_k(G)$ is finite-dimensional over C [10, p.46].

Let B_k be a C-basis for $A_k(G)$, and let

$$B = \bigcup_k B_k.$$

B is called a graded basis for $A(G)$, and Q_B denotes the field generated by all the $a_n(f)$, where f ranges over B and n ranges over the non negative integers.

Theorem.

$$\bigcap Q_B \,{}_{\text{B a graded basis}} = Q_{\tilde{B}}$$

for some graded basis $\tilde{B}$ *and* $Q_{\tilde{B}}$ *is a finitely generated field.*

Proof. Fix an even $k \geq 0$, and let $B_k = \{f_1,...,f_m\}$, where $m = m(k)$. We then have an $m \times \infty$ matrix

$$\begin{pmatrix} a_0(f_1) & a_1(f_1) & \dots & a_n(f_1)\dots \\ a_0(f_2) & a_1(f_2) & \dots & a_n(f_2)\dots \\ & & \cdot & \\ & & \cdot & \\ a_0(f_m) & a_1(f_m) & \dots & a_n(f_m)\dots \end{pmatrix}.$$

This matrix has rank m, and so there are integers $d_1,...,d_m$, $0 \leq d_1 < ... < d_m$ such that if $a_{ij} = a_{d_j}(f_i)$, then $(a_{ij})_{1 \leq i,j \leq m}$ is non singular. If (c_{ij}) is the inverse matrix, and we set

$$\tilde{f}_i = c_{i1}f_1 + \dots + c_{im}f_m,$$

then $\tilde{B}_k = (\tilde{f}_1,..., \tilde{f}_m\}$ is a basis for $A_k(G)$, and $a_{d_j}(\tilde{f}_i) = \delta_{ij}$. If now $\{f_1^*,...,f_m^*\}$ is any basis for $A_k(G)$, then $f_i^* = u_{i1}\tilde{f}_1 + ... + u_{im}\tilde{f}_m$, and $a_{d_j}(f_i^*) = u_{i1}a_{d_j}(\tilde{f}_1) + ...$ $u_{im}a_{d_j}(\tilde{f}_m) = u_{ij}$. Thus, the field F generated by all the $a_n(f_i^*)$ contains all the u_{ij}.

Since each $\tilde{f}$ is a $Q(\{u_{ij}\})$-linear combination of the f^*, it follows that F coincides with the field generated by all the u_{ij} and all the $a_n(\tilde{f}_j)$. Thus, if we set $\tilde{B} = \cup_k \tilde{B}_k$, then the first assertion of the theorem holds. It remains to show that $Q_{\tilde{B}}$ is finitely generated.

Since G has genus zero, there is a covering map $\lambda : \hat{h} \to \hat{C}$ such that λ induces an analytic bijection between $G\backslash\hat{h}$ and $\hat{C}$; if λ, $\tilde{\lambda}$ are two such maps, then $\tilde{\lambda} = \frac{a\lambda+b}{c\lambda+d}$ for some $\begin{pmatrix} a & b \\ c & d \end{pmatrix} \in GL_2(C)$. Set $p_i = \lambda(q_i)$, $1 \le i \le r$, $S = \{p_1,\ldots,p_r\}$. From [12], it follows that there is a map $c = c_S : S \cap C \to C$ such that for $\tau \in \hat{h}$,

$$\frac{\lambda'''(\tau)}{\lambda'(\tau)^3} - \frac{3}{2}\frac{\lambda''(\tau)^2}{\lambda'(\tau)^4} + \sum_{s \in S \cap C} \left\{ \frac{1}{2}(\lambda(\tau) - s)^{-2} + c(s)(\lambda(\tau) - s)^{-1} \right\} = 0.$$

We assume without loss of generality that $S \subseteq C$. Then $\lambda = p_1 + \sum_{n=1}^{\infty} a_n(\lambda)e_1^n$, and the preceding equation gives us a Laurent series in e_1, from which we deduce easily that

$$Q(p_1,a_1(\lambda),\ldots,a_n(\lambda),\ldots) \subseteq Q(S, \text{im } c_S, a_1(\lambda)). \qquad (2)$$

Set $\mu = \frac{1}{2\pi i}\lambda' = \sum_{n=1}^{\infty} na_n(\lambda)e_1^n$. Then each element of $A_k(G)$, $k = 2l$, is of the form $F \cdot \mu^l$, $F \in C(\lambda)$; more precisely,

$$F = \frac{H(\lambda)}{(\Pi_{s \in S}(\lambda - \lambda(s)))^{l}}, \qquad H(\lambda) \in C[\lambda].$$

The only restraint on $H(\lambda)$ is that $\deg H(\lambda) \le rl - 2l$, which inequality guarantees that $F\mu^l$ is holomorphic at the poles of λ (each of which, as $S \subseteq C$, is in h). So $\dim_C A_{2l}(G) = rl - 2l + 1$. This in turn implies that A(G) is generated by 1 and $A_2(G)$. Moreover, $\{1\} \cup \left\{ \frac{\lambda^i}{\Pi_{s \in S}(\lambda - \lambda(s))^l} \cdot \mu \mid 0 \le i \le r-2 \right\}$ generates a subring of A(G) which contains a C-basis, so by inspection, (2) implies the theorem.

The field $\mathbf{Q}_{\mathfrak{B}}$ depends on G and on G(∞), but not on the particular element of G(∞) which we choose to build Fourier coefficients, so we get, for each i, a well defined $\mathbf{Q}_{G,G(q_i)}$, the field generated by the Fourier coefficients of a suitable basis of A(G), by using a suitable local variable at q_i.

The case r = 4 is exceptional. In this case, G is contained in a group $\tilde{G} \in \mathcal{F}(0; 1; 2,2,2)$ [11]. For $\tilde{G}$, we get a well-defined field $\mathbf{Q}_{\tilde{G}}$, since $\tilde{G}$ has just one cusp. These fields $\mathbf{Q}_{G,G(q_i)}$, $\mathbf{Q}_{\tilde{G}}$ are somewhat strange, it seems to me. My conjecture is that if $\lambda(\{q_1,\dots,q_r\}) \subseteq \bar{\mathbf{Q}} \cup \{\infty\}$, then $\mathbf{Q}_{G,G(q_i)} \subseteq \bar{\mathbf{Q}}$. The Chudnovsky brothers, in unpublished work, have taken the case r = 4, $\lambda(\{q_1,\dots,q_4\}) = \{\infty,0,1,n\}$, chosen particular integers n, and have shown that for their choices no element of im c_s is the root of a non-zero polynomial in one variable with integer coefficients whose degree is ≤ 100 and whose coefficients are ≤ 10^{100} in absolute value, giving strong evidence that im c_s consists of transcendental numbers. I yield to this superior display of approximation-theoretic techniques, but .... .

References

1. A Beardon, *The Geometry of Discrete Groups* (Springer-Verlag, Graduate Texts in Mathematics **91**, 1983).
2. G V Belyi, On Galois Extensions of a maximal cyclotomic field, *Math. USSR Izvestija*,**14** (1980), 247-256.
3. C Curtis and I Reiner, *Methods of Representation Theory Vol II*, (John Wiley & Sons, 1987).
4. R Guralnick and J G Thompson, Groups of genus zero, to appear.
5. D Hilbert, *Gesammelte Abhandlungen*, (Chelsea, New York, 1965).
6. H Petersson, Zur analytischen Theorie der Grenzkreisgruppen, I, II, III, IV, V, *Math. Ann.* **115** (1938), 23-67, 175-204, 518-572, 670-709, *Math. Z.* **44** (1939), 127-155.
7. B Schoeneberg, *Elliptic Modular Functions*, (Springer-Verlag, Die Grundlehren der Mathematischen Wissenschaften in Einzeldarstellungen, **203**, 1974).
8. J-P Serre, *A Course in Arithmetic*, (Springer-Verlag, Graduate Texts in Mathematics **7**, 1970).
9. J-P Serre, *Topics in Galois Theory*, (Harvard Lecture Notes (written by H Darmon), 1988).
10. G Shimura, *Introduction to the Arithmetic Theory of Automorphic Functions*, (Princeton University Press, 1971).

11. D Singerman, Finitely maximal Fuchsian groups, *J. London Math. Soc.* (2), **6** (1972), 29-38.

12. J G Thompson, Algebraic numbers associated to certain punctured spheres, *J. Algebra* **104** (1986), 61-73.

13. J G Thompson, Some finite groups which appear as Gal L/K, where K $\subseteq$ Q(μ_n), *J. Algebra* **89** (1984), 437-499.

INFINITE SIMPLE PERMUTATION GROUPS - A SURVEY

J K TRUSS

University of Leeds, Leeds LS2 9JT

1. Introduction

Simple groups arise naturally as building blocks for an arbitrary group. The success of the classification of the finite simple groups as a major step in this programme is however hardly mirrored in the infinite case, and here the results about simple groups, and the examples we know of, while constantly being added to, are far from complete and seem destined to remain so. One feature of the finite simple groups which we may focus on is that they usually admit a permutation representation (often more than one) of a particularly beautiful or "homogeneous" structure of some sort. We can analyse the structures (which may be graphs, Steiner systems, projective spaces etc.) hand-in-hand with their groups, and the understanding of each should enrich the study of the other.

In the case of infinite permutation groups, consideration of automorphism groups of "symmetrical" structures often provides natural examples of simple groups, since the homogeneity of the structure may ensure that the group has only few normal subgroups - or if many we may still be able to describe what they are. In this paper I shall survey results of this sort and indicate in what way I hope that they will assist a study of the structures concerned.

Firstly let us try to pin down what we mean by "structure". Usually this will mean "countable relational structure

$$\mathcal{A} = < A : \{R_\lambda\}_{\lambda \in \Lambda} >".$$

Here A is a set of cardinality $\aleph_0$, and each R_λ is an n-ary relation on A for some finite n = 1, 2, 3,.... We do not always observe the cardinality restriction on A, and indeed the only reason for mentioning it here is that most of the interesting examples we have in mind *are* countable. From the point of view of the permutation group arising however there is no loss of generality in supposing as we have that the structure is *relational*, i.e. does not have functions or constants named. For if $\mathcal{A}$ is a structure which may have functions or constants then we can associate a relational structure $\mathcal{A}'$ with it having the same domain and automorphism group; this is done by replacing each n-ary f by an n+1-ary relation

$$R_f = \{(x_1,...,x_{n+1}) : f(x_1,...,x_n) = x_{n+1}\}$$

and each constant c by a unary relation $R_c = \{c\}$. As well as structures which are initially presented to us as relational, such as partial orderings, graphs, digraphs, and geometries, we may therefore also handle for example automorphism groups of groups, vector spaces, and algebras. It will certainly be important to consider such cases from the point of view of the general theory, though most of the examples we discuss here will be given in relational form from the word go.

In addition to structures in the strict (model-theoretic) sense it may also be fruitful to consider other kinds of mathematical objects, of which the prime example for us is a topological space. Here one looks at the *homeomorphisms* of the space to itself rather than the *automorphisms*. The resulting group is often interesting in its own right, having a rich and beautiful structure. This is particularly true when the space is the set $\mathbb{Q}$ of rational numbers under the usual topology, and this group Aut $\mathbb{Q}$ has been studied on its own merits in [3, 17] and in its relationship to other permutation groups in [16, 24]. To enlarge the class of topological spaces under consideration one really needs to relax the cardinality restriction and look at interesting examples like Cantor space 2^ω and Baire space ω^ω as in [26]. More general results were proved by Anderson in [1].

The class of automorphism groups of relational structures may be singled out from the class of all permutation groups by characterizing them as the closed subgroups of the full symmetric group under an appropriate topology. If the underlying set Ω is countable,

$$\Omega = \{x_n : n \in \mathbb{N}\}$$

say, then the topology may be induced by the metric d where

$$d(s,t) = \Sigma\{2^{-n} : \sigma x_n \neq \tau x_n \text{ or } \sigma^{-1}x_n \neq \tau^{-1}x_n\}.$$

This also allows us to assign a "complexity" to various other permutation groups; for example the group AAut Γ studied in [25] is F_σ, and Aut $\mathbb{Q}$ is Π^1_1 (and probably complete Π^1_1). It would be much more satisfactory however to be able to view a greater variety of permutation groups as "closed" in some sense, since closed groups are generally much easier to handle, and it is then often possible to apply the notions of Baire category. We shall give an example in Section 5 of how this can be done by defining an alternative metric on Aut $\mathbb{Q}$ reflecting its natural approximation structure as discussed in [24]. It is hoped that this approach - of considering more general metrics or topologies than the "obvious" ones - will enable one to carry out an analysis of a larger class of permutation groups than just the automorphism groups of relational structures.

I would like to thank Matatyahu Rubin for interesting discussions and for allowing me to quote the results of [18, 19] in this paper.

2. Groups of order-preserving permutations

A good starting-point is work of Higman [11] who considered the case of a linearly ordered set $(\Omega,\leq)$ and its automorphism group

$$\mathrm{Aut}(\Omega,\leq) = A(\Omega)$$

(in the notation of [8]). There is a wealth of information about this case (see [8] and its bibliography) so we shall only comment on a few aspects. Firstly we shall assume that the group acts 2-homogeneously on Ω, i.e. transitively on the family of unordered pairs $\{x,y\}$ of members of Ω. It is clear that $A(\Omega)$ has at least three proper non-trivial normal subgroups, written

$$L(\Omega) = \{\sigma \in A(\Omega) : \mathrm{supp}\ \sigma \text{ bounded above}\},$$

$$R(\Omega) = \{\sigma \in A(\Omega) : \mathrm{supp}\ \sigma \text{ bounded below}\},$$

and

$$B(\Omega) = \{\sigma \in A(\Omega) : \mathrm{supp}\ \sigma \text{ bounded above and below}\},$$

see Fig. 1.

(Here supp σ is the set of elements of Ω moved by σ.)

The hypothesis of 2-homogeneity is sufficient to ensure that these groups are all proper, non-trivial, and distinct, and Higman showed (under the same assumption) that $B(\Omega)$ is simple and so there is no non-trivial normal subgroup of $A(\Omega)$ contained in it. This example raises a number of interesting questions, of which we mention just four.

(2.1) If $\sigma,\tau \in A(\Omega)$ and τ lies in the normal subgroup of $A(\Omega)$ generated by σ, is there a bound on the number of conjugates of σ and σ^{-1} required in an expression for τ?

(2.2) Are $L(\Omega)$, $R(\Omega)$, and $B(\Omega)$ the only proper non-trivial normal subgroups of $A(\Omega)$?

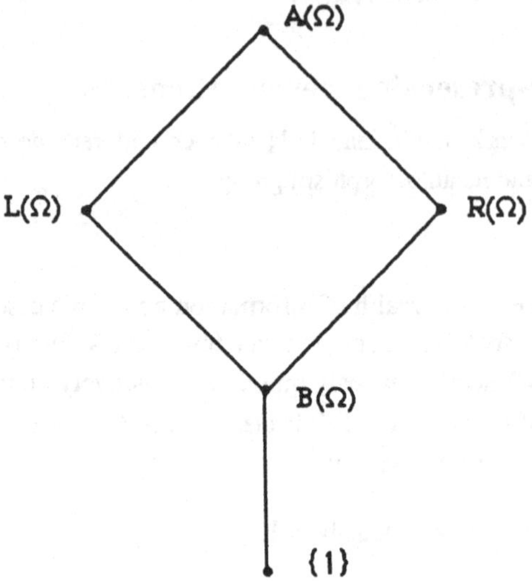

Fig.1

(2.3) Higman's example provides us with three groups corresponding to each doubly homogeneous linear ordering,

B(Ω), L(Ω)/B(Ω), and R(Ω)/B(Ω).

(There are trivial isomorphisms A(Ω)/L(Ω) ≅ R(Ω)/B(Ω) and A(Ω)/R(Ω) ≅ L(Ω)/B(Ω)). In some cases all three of these are simple, for example if $\Omega = \mathbf{Q}$ or **R**. How many examples of infinite simple groups does this really give us? Are these three distinct, and what happens for different values of Ω?

(2.4) What information can be recovered from the *abstract group* A(Ω) (as opposed to A(Ω) viewed as *permutation group*) about the structure (Ω, ≤), using for example answers to the previous three questions?

Partial answers have been found to all these questions, but more remains to be done.

(2.1) This question was explored in [2] and [26] and those results were extended in [6]. Let us illustrate in an easy case, namely for B(Ω). Here if $\sigma,\tau \neq 1$ then one can show that τ is the product of a conjugate of σ with a conjugate of σ^{-1}, so 2 is the optimum number in this case. For a proof let supp $\tau \subseteq (a,b)$, and let x be such that $\sigma x \neq x$. Without loss of generality suppose $x < \sigma x$. By double homogeneity there is $\theta \in B(\Omega)$ such that

$$x < \theta a < \theta b < \sigma x.$$

Thus

$$\text{supp } \theta\tau\theta^{-1} \subseteq (x,\sigma x)$$

and from this we deduce (essentially by [8, Theorem 2.2.5]) that σ and $\sigma\theta\tau\theta^{-1}$ are conjugate, from which the result follows.

It was shown in [26] that the minimum number of conjugates in general required in $A(\mathbf{Q})$ is 4, and this is extended in [6] to $A(\Omega)$ for a wide class of other values of Ω (specifically all 2- homogeneous Ω), including for example

$$\mathbf{R}, \mathbf{Q} \times \omega_1 \text{ and } \overline{\mathbf{Q} \times \omega_1} \quad \text{(the "long line").}$$

The proofs are however much more involved.

(2.2) Here the answer is that these are the only proper non-trivial normal subgroups of $A(\Omega)$ under only very restricted circumstances, namely when Ω contains a countable subset which is unbounded above and below in Ω. A complete construction of all normal subgroup lattices of $A(\Omega)$ for doubly homogeneous Ω was given in [2, 5], and many further properties of the lattice were derived in [4]. The construction employs some ideas from set theory; we illustrate the ideas in one case by producing a different normal subgroup of

$$A(\overline{\mathbf{Q} \times \omega_1})$$

where $\overline{\mathbf{Q} \times \omega_1}$ is the *long line*, i.e. the order-completion of ω_1 copies of $\mathbf{Q}$ (so-called because all of its proper initial segments are order-isomorphic to $\mathbf{R}$).

Let us say that the subset X of Ω is *closed unbounded* if (i) X is closed in the order topology on Ω, and (ii) X is unbounded in Ω. Let $\mathcal{F}$ be the family of closed unbounded sets in $\Omega = \overline{\mathbf{Q} \times \omega_1}$ of the form

$$\bigcup_{\alpha \in \omega_1} [a_\alpha, b_\alpha]$$

where

$$a_0 < b_0 < a_1 < b_1 < \ldots .$$

To be closed, any such expression must of course fulfil $a_\lambda = \sup_{\alpha<\lambda} a_\alpha$ for all limit ordinals $\lambda < \omega_1$. For $\sigma \in A(\Omega)$ we let fix σ be the set of fixed points of σ (= the complement of supp σ) and we let

$$N = \{\sigma \in A(\Omega) : \text{fix } \sigma \text{ contains a member of } \mathcal{F}\}.$$

Then N is the desired normal subgroup.

Obviously N is closed under inverses and conjugacy, and $N \nleq L(\Omega), R(\Omega)$ since we may define $\sigma \in N - (L(\Omega) \cup R(\Omega))$ by

$$\sigma(x,\alpha) = \begin{cases} (x,\alpha) & \text{if } \alpha \text{ is a limit ordinal} \\ (x+1,\alpha) & \text{otherwise.} \end{cases}$$

On the other hand it is equally clear that $N \neq A(\Omega)$. Showing that N is a subgroup amounts to checking that $\mathcal{F}$ is closed under finite intersections. To see that this holds let $X = \bigcup_{\alpha<\omega_1} [a_\alpha,b_\alpha]$ and $Y = \bigcup_{\alpha<\omega_1} [c_\alpha,d_\alpha] \in \mathcal{F}$. We choose

$$x_0 < y_0 < x_1 < y_1 < \dots$$

by transfinite induction with $x_\lambda = \sup_{\alpha<\lambda} x_\alpha$ for limit ordinals λ and

$$[x_\alpha,y_\alpha] \subseteq X \cap Y.$$

Assume x_α and y_α have been defined, for some $\alpha < \omega_1$. We choose ordinals $\gamma_0 < \gamma_1 < \dots$ and $\delta_0 < \delta_1 < \dots$ such that

$$y_\alpha < a_{\gamma_0} < c_{\delta_0} < a_{\gamma_1} < c_{\delta_1} < \dots$$

and we let

$$x_{\alpha+1} = \sup_{n\in\omega} a_{\gamma_n} = \sup_{n\in\omega} c_{\delta_n}.$$

Then if $\lambda = \sup_{n\in\omega} \gamma_n$ and $\mu = \sup_{n\in\omega} \delta_n$, $x_{\alpha+1} = a_\lambda = c_\mu$ so if

$$y_{\alpha+1} = \min(b_\lambda,d_\mu)$$

then $[x_{\alpha+1},y_{\alpha+1}] \subseteq X \cap Y$. The cases of zero and limit ordinals are treated quite easily.

It turns out that this N is the unique minimal normal subgroup of $A(\Omega)$ properly containing $L(\Omega)$. The full picture as to what the normal subgroups are requires an analysis of $\mathcal{P}(\omega_1)$ modulo the ideal of "non-stationary" sets (those disjoint from some closed unbounded set) .

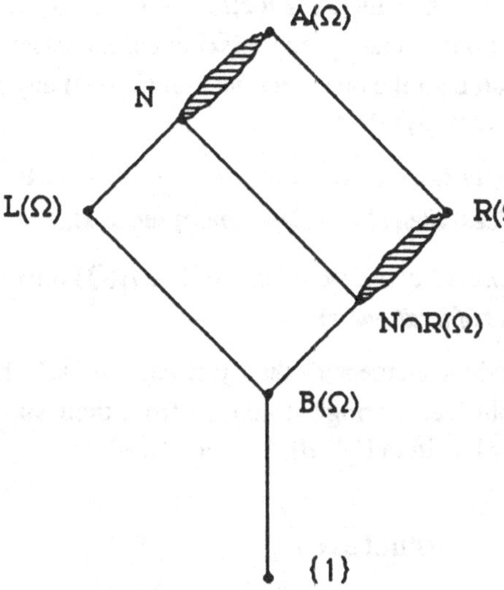

(2.3) A technique was developed in [26] for distinguishing these groups in some cases. For example among the groups arising when

$$\Omega = \mathbf{Q}, \mathbf{Q} \times \omega_1, \mathbf{Q} \times (\omega_1{}^*+\omega_1)$$

or their order-completions (where $\omega_1{}^*$ is ω_1 with the reverse ordering), the method is capable of distinguishing between the groups in all possible cases. By "possible" we mean unless they are (obviously) isomorphic, like $L(\mathbf{Q}) \cong R(\mathbf{Q})$ etc. Moreover the method tells us something about the *elementary* theory of the groups, i.e. about their properties expressible just by quantifying over elements of the group (as opposed to subsets, subgroups, etc.). To illustrate we sketch how one can establish $A(\mathbf{Q}) \not\equiv A(\mathbf{R})$, - distinguishing these groups by an elementary property.

Let $\varphi(\sigma)$ be the formula

$$(\forall\theta)(\exists\alpha)(\theta^{-1}\sigma\theta\sigma = \alpha^{-1}\sigma\alpha).$$

This says that the product of σ with a conjugate of σ is itself also necessarily conjugate to σ. An example of such an element in $A(\mathbf{Q})$ is given by $\sigma x = x+1$, all x, translation by 1 to the right. One can show that in $A(\mathbf{R})$ there are just 7 elements (up to conjugacy) satisfying $\varphi(\sigma)$, but in $A(\mathbf{Q})$ there are 11 and these two facts may be easily written down as elementary formulae. The difference relies on the existence of elements satisfying $\varphi(\sigma)$ whose support is of the form (a,∞) or of the form $(-\infty,a)$ where a may be rational or irrational.

(2.4) Gurevich and Holland gave in [9] a method for "reconstructing" the underlying set Ω from its automorphism group $A(\Omega)$ in certain cases. Of course this gives a much better result than the one mentioned in (2.3), at any rate for the cases of **Q** and **R**. Their theorem says the following:

(i) *if* Ω *is a doubly homogeneous chain such that* $A(\Omega) \equiv A(\mathbf{R})$ (*actually elementary equivalence suffices*) *then* Ω *is order-isomorphic to* **R**,

(ii) *if* Ω *is a doubly homogeneous chain such that* $A(\Omega) \equiv A(\mathbf{Q})$ *then* Ω *is order-isomorphic to* **Q** *or* **Ir** (*the set of irrationals*).

Giraudet has further explored the elementary theory of $A(\Omega)$ in [12, 13, 14], and other important results about "recovering" structures from their automorphism groups have been derived by Rubin in [19, 20, 21] and elsewhere.

3. Other homogeneous structures

The groups discussed in Section 2 are all torsion-free. The "canonical example" $A(\mathbf{Q})$ of the behaviour we wished to discuss arises from a particularly homogeneous structure $(\mathbf{Q},<)$. Informally this means that there are many parts of the structure which look alike; formally this is captured by saying that a structure $\mathcal{A}$ is *homogeneous* if any isomorphism between finite substructures extends to an automorphism of $\mathcal{A}$. This will generally mean that Aut $\mathcal{A}$ is "large", specifically that it has cardinality $2^{\aleph_0}$ (where $\mathcal{A}$ is taken to be countable), though this is not always the case; Aut$(\mathbf{Q},+,0)$ is countable (isomorphic to $(\mathbf{Q}-\{0\},\times)$ in fact) and we may regard $(\mathbf{Q},+,0)$ as homogeneous provided we relax our previous restriction to relational structures and interpret "finite substructure" as meaning "finitely generated substructure".

With this in mind we move on to consider various other types of homogeneous structure, whose automorphism groups contain elements of finite order. A first and familiar example is that of Symm Ω itself (where the "structure" is trivial). Here the group is *not* simple, but its normal subgroups are well known, and are just $\mathrm{Symm}_{<\omega}\,\Omega$ (the finitary permutation group) and Alt Ω (the alternating group). As a generalization of this we may consider homogeneous graphs on a countable set Γ. These were classified by Lachlan and Woodrow [15] and remarkably there are only countably many, up to isomorphism. Many of them arise by rather straightforward constructions from complete graphs, so that the resulting automorphism groups turn out to be wreath products of symmetric groups etc. Some of the structures are however of great independent interest, and are "primitive" - meaning that their automorphism groups are primitive permutation groups. Among these are Rado's universal graph Γ_2 and as generalizations the K_n-free universal homogeneous graphs for n = 3,4,5,... Among many equivalent constructions the simplest characterization of Γ_2 is this: $|\Gamma_2| = \aleph_0$ and if U,V are disjoint finite sets of vertices then there is

$$x \in \Gamma_2 - (U \cup V)$$

joined to all members of U and to none of V. With this definition it turns out that such a Γ_2 exists, and is unique up to isomorphism, and has an automorphism group of cardinality $2^{\aleph_0}$ and a very rich structure.

A detailed study of Aut Γ_2 and certain generalizations, Aut Γ_C for C a set of "colours" with $2 \leq |C| \leq \aleph_0$ was carried out in [23] (here it is the *edges* which are coloured, not the vertices). The main result was that each Aut Γ_C is a simple group. In fact it was shown that if $\sigma,\tau \neq 1$ then τ is the product of 5 conjugates of σ.

Now in many simplicity proofs the following outline is adopted. A certain family Σ of "canonical" members of the group G is chosen. One then shows

(1) that if $1 < N \lhd G$ then $N \cap \Sigma \neq \emptyset$,

(2) Σ generates G.

For example if $G = \mathcal{A}_n$, one often takes Σ to be the family of 3-cycles. In the case of $B(\Omega)$ considered by Higman one may take Σ to be the family of all elements with a single increasing "orbital" (= an open interval spanned by an orbit). For Aut Γ_2 in [23] we took for Σ the family of elements of cycle type ∞^∞ together with an additional "universality" property. It is not clear that this is the best choice, and may be responsible for providing the non-optimal number of 5 conjugates. Rubin has shown that 4 is possible and that 3 conjugates are necessary, so the precise best number is still unknown. We make some remarks on this in Section 5.

In view of the fact that the proof of the simplicity of Aut Γ_C appeared to rely just on very general homogeneity properties of Γ_C it seemed clear that the result should apply much more widely. In particular the following were "test cases" for simplicity:

(3.1) $(\text{Aut}\ \Gamma_C)_{(X)}$, the pointwise stabilizer of a finite subset X of Γ_C,

(3.2) Aut $\mathcal{T}$, where $\mathcal{T}$ is the countable universal homogeneous tournament,

(3.3) Aut $\Gamma(n)$, where $\Gamma(n)$ is the countable universal K_n-free graph.

In [18] Rubin succeeded in isolating sufficient conditions on a structure for it to have a simple automorphism group, and these conditions applied to all three test cases and to many others. Although the conditions are too technical to give in detail we mention some points about them. Firstly the structures are all obtained by a canonical method for producing countable homogeneous structures due to Fraïssé [7]. This works for a class $\mathbf{C}$ of finite relational structures provided it is closed under isomorphism, substructures, and has the amalgamation property. If for example $\mathbf{C}$ consists of the class of all finite linear orderings then the resulting homogeneous structure is isomorphic to $(\mathbb{Q},<)$, or if $\mathbf{C}$ is the class of all finite graphs then the resulting homogeneous structure is the random graph. Rubin's structures, which he calls "simple" are thus characterized by certain amalgamation classes $\mathbf{C}$ of finite structures. The second point is that his structures are all binary, that is to say all the relations occurring are 1- or 2-place. The main extra condition imposed on $\mathbf{C}$ is that the amalgamations may be carefully controlled in terms of the binary relations available. We now quote Rubin's main theorem, and some of its consequences.

Theorem 3.1. *If $\mathcal{A}$ is simple and non-trivial then* Aut $\mathcal{A}$ *is simple and moreover for any non-identity* $\sigma,\tau \in$ Aut $\mathcal{A}$, *τ is the product of* 5 *conjugates of* σ.

We make some remarks in Section 5 about the number 5 of conjugates required.

Looking at the case of homogeneous graphs Rubin was able to show that all the primitive graphs in Lachlan and Woodrow's list are simple, so the following result is immediate.

Theorem 3.2. *Let Γ be countable homogeneous graph. If* Aut Γ *is primitive, then it is simple.*

As far as stabilizers are concerned, Rubin showed not only that $(\mathrm{Aut}\ \Gamma_C)_{(X)}$ is a simple group for finite $X \subseteq \Gamma_C$, but also that the class of automorphism groups of simple structures is closed under formation of such stabilizers.

Theorem 3.3. *Let $\mathcal{A}$ be a simple structure. Then for every finite* $X \subseteq A$, *there is a simple structure having domain* $A - X$ *and automorphism group* $(\mathrm{Aut}\ \mathcal{A})_{(X)}$ *(where this is strictly speaking the permutation group induced in the natural way on* $A - X$*).*

We remark that from this case one can see why it is necessary to allow extra unary and binary relations in a "simple" structure. For instance if Γ_2 is the random graph, and $|X| = n$, then X naturally subdivides $\Gamma_2 - X$ into 2^n pieces which must be fixed by $(\mathrm{Aut}\ \Gamma_2)_{(X)}$, corresponding to the possible combinations of adjacencies with elements of X. Moreover it is certainly not possible to allow the automorphism groups of these pieces to act independently as there are many adjacencies and non-adjacencies between their members which have to be preserved. The definition of simple entails that there will be binary relations which "keep track of" the relations between these pieces.

So far we have only indicated how to find countably many simple permutation groups. Henson [10] gave a method for constructing $2^{\aleph_0}$ pairwise non-isomorphic countable homogeneous directed graphs, and these are all simple, so that Theorem 3.1 applies, and we have continuum many examples. A snag here is that it is not at all obvious that just because two structures are non-isomorphic, they have non-isomorphic automorphism groups, and this is similar to the issue raised at the end of Section 2.

To take a case in point, it was shown in [23] that if

$$2 \le |C_1| < |C_2| \le \aleph_0$$

then Aut Γ_{C_1} and Aut Γ_{C_2} are non-isomorphic as permutation groups, but it was left unresolved as to whether they might be isomorphic as abstract groups. This was settled in [19] where it was shown how to recover enough of the original structure of Γ_{C_1} or Γ_{C_2} from the group structure alone, and hence to be able to distinguish them up to isomorphism. It still seems to be open whether or not Aut Γ_{C_1} and Aut Γ_{C_2} are elementarily equivalent.

4. Topological results

The idea which enabled Higman to prove the simplicity of B(Q) was the "homogeneity" of the structure, which we take here in an informal sense, just meaning that there are many pieces of the ordered set (in this case bounded open intervals) which look alike. Although the most extensive applications of this have been in the study of order-preserving permutation groups, another natural direction to follow is that of topological spaces where many of the open sets are homeomorphic via homeomorphisms of the whole space ("internally" as it were). This was looked at by Anderson in [1] where he established the simplicity of the group Aut X of homeomorphisms to itself of a topological space X for various possible values of X.

Anderson worked with a family $\mathcal{K}$ of closed subsets of X any two of which are homeomorphic, such that any non-empty open set contains a member of $\mathcal{K}$ and

$$Y \in \mathcal{K} \Rightarrow \overline{X - Y} \in \mathcal{K}.$$

If such a family exists then X is said to be $\mathcal{K}$ *-structured.* For zero-dimensional spaces such as **Q**, **Ir**, and 2^{ω} it is easiest to take for $\mathcal{K}$ the family of proper non-empty clopen subsets, but for the sphere S^2 for example one takes $\mathcal{K}$ to be the family of closed 2-cells. (Note that boundedness considerations imply that $\mathbf{R}^n$ is not $\mathcal{K}$ -structured for any $n \geq 1$ or $\mathcal{K}$; here Aut X has a (proper) minimal normal subgroup, the group of homeomorphisms of bounded support.) In addition Anderson defines what it means for X to be (G, $\mathcal{K}$) - *homogeneous* where G ≤ Aut X. In essence what is required here is that it should be possible to move the members of $\mathcal{K}$ sufficiently freely using members of G to avoid each other, and that any member of $\mathcal{K}$ should contain a sequence of pairwise disjoint members of $\mathcal{K}$ on which appropriate actions of members of G can be combined; this is a "piecewise patching" type of condition, also used heavily in the order-preserving case (see [6]).

Theorem 4.1. *Let* X *be* (G, $\mathcal{K}$)*-homogeneous and let* $\sigma, \tau \in G$, $\sigma \neq 1$. *Then* τ *is the product of six conjugates of* σ *(written alternately). Hence* G *is simple.*

Note that this result of Anderson's applies not just to Aut X, but also to sufficiently large subgroups.

In [26] we looked at this result and tried to reduce the number of conjugates required. This met with only limited success. That is to say in certain cases, for example X = **Q** or **Ir**, the number was reduced to "three" and this was shown to be optimal. Undoubtedly this should apply much more widely. A surprisingly troublesome case is Cantor space 2^{ω}, and the snag is essentially its compactness. As Anderson's conditions suggest, many of the arguments rely on being able to "cut up" the members of $\mathcal{K}$ into countably many pieces and patch the effect of possibly different homeomorphisms on the pieces. If X is compact then so is each member of $\mathcal{K}$ so no $Y \in \mathcal{K}$ can be an infinite disjoint union of members of $\mathcal{K}$. To modify the arguments to apply to this case one needs to allow the pieces to form a partition of a *subset* of Y rather than the whole of Y and it is also then necessary to take more care when "patching" to ensure that the resulting map is continuous. Strangely enough, in analysing the stabilizer of a singleton in Aut 2^{ω}, as we did in [26], things become a little easier, and this is essentially because the space resulting from the removal of a point is no longer compact, and infinite partitions into pieces which look alike are now possible. We conclude this section by quoting the result from [26] about pointwise stabilizers in Aut **Q** of finite subsets of **Q** where considerations of this sort arise.

Theorem 4.2. *Let* A *be a finite subset of* **Q**. *Then the non-identity normal subgroups of the pointwise stabilizer* (Aut **Q**)$_{(A)}$ *of* A *in* Aut **Q** *are precisely those subgroups of* (Aut **Q**)$_{(A)}$ *of the form*

$$\{\sigma \in (\mathrm{Aut}\,\mathbf{Q})_{(A)} : (\forall b \in B)\ \sigma \textit{ fixes pointwise a neighbourhood of } b\}$$

for subsets B *of* A.

Suppose for example that A has just a single point a. Then B may be Ø or {a} in which case the subgroups in question are (Aut **Q**)$_a$ and

$$L(a) = \{\sigma : \sigma \text{ fixes pointwise a neighbourhood of } a\}$$

respectively. Since L(a) is the union of a chain of length ω of groups isomorphic to Aut **Q**, its simplicity is easy to check. The main part of the proof consists in showing that if a normal subgroup N of (Aut **Q**)$_a$ contains an element moving points arbitrarily close to a, then

$$N = (\mathrm{Aut}\,\mathbf{Q})_a.$$

The key idea here is to adapt Anderson's arguments by taking for $\mathcal{K}$ the family of all open subsets Y of **Q** - {a} such that Y and **Q** - Y - {a} each "abut" a, meaning that

$$\overline{Y} = Y \cup \{a\}$$

and

$$\overline{\mathbf{Q} - Y - \{a\}} = \mathbf{Q} - Y.$$

The fact that three conjugates suffice is then shown in a similar fashion to Aut **Q**.

5. Generic elements and the least number of conjugates

Suppose we select an element of the permutation group G on the countable set Ω at random. What is it most likely to be? Can we in some way describe the element which arises in terms of its cycle or other structure? Also in cases where G is simple, can we be sure that such elements form the "intermediate stage " in some (natural) proof of the simplicity for G (i.e. the elements of the family Σ in the notation of Section 3)? In cases where G is not simple the next best thing would be for G to be the normal closure of any such element.

In fact the word we use to describe the type of element we have in mind is *generic* (by analogy with the notion in set theory and model theory) and the framework for formulating the correct definition will be topological. Intuitively to say that an element σ of G is generic will mean that any property of σ which can be established on the basis of finitely many of its values and which is "consistent" with σ actually holds. To make this idea more precise we suppose that G has been endowed with a complete metric d. The principal example we have in mind is where G = Aut $\mathcal{A}$ is the automorphism group of some relational structure, in which case the natural metric to take (and the one pertinent to discussion of "finite pieces of information") is given by

$$d(\sigma,\tau) = \Sigma\{2^{-n} : \sigma x_n \neq \tau x_n \text{ or } \sigma^{-1}x_n \neq \tau^{-1}x_n\}$$

where $\{x_n : n \in \omega\}$ is an enumeration of Ω. We refer to this as the *standard* metric.

Since the metric is supposed to be complete, the Baire category theorem applies and we may meaningfully talk about sets of the first category, or *meagre* sets, and their complements, *comeagre* sets. Relative to such a metric d on G we then define $\sigma \in$ G to be *generic* if it lies in a comeagre conjugacy class. If the specific metric is not mentioned then we assume it is the standard one. Now it is not guaranteed that there will be *any* generic elements at all. For instance there are three other clear possibilities. One is that all conjugacy classes are meagre; the second is that some conjugacy class fails to have the "property of Baire". (A set is said to have the *property of Baire* if it differs from an open set by a meagre set). The third is that some conjugacy class is comeagre on a non-empty open set but not on the whole space.

Further remarks will be made about these cases in [28]. With regard to the second, examples of this will have to be constructed in a rather "unnatural" way, since any Borel set has the property of Baire and in the most obvious examples, like Symm Ω and Aut $\mathcal{A}$ for various homogeneous $\mathcal{A}$, all conjugacy classes are Borel. (Moreover by [22] it is relatively consistent with ZF set theory that *all* sets have the property of Baire). The third possibility can occur, but not if $\mathcal{A}$ has sufficient homogeneity. As an example let $\mathcal{A}$ be a graph with domain Z and edges $\{n,n+1\}$ for each n. Then any automorphism is conjugate to a reflection or translation, so in Aut $\mathcal{A}$ (the infinite dihedral group) there are only countably many conjugacy classes, all of which are open. Similar remarks apply when $\mathcal{A}$ is a tree in which every vertex has degree n, fixed $n \geq 2$.

Let P be the family of finite partial automorphisms of $\mathcal{A}$, partially ordered by extension, and for $p \in$ P let

$[p] = \{\sigma \in \text{Aut}\,\mathcal{A} : \sigma \text{ extends } p\}.$

Then saying $\mathcal{A}$ is homogeneous just means that each [p] is non-empty, and in any case $\{[p] : p \in P\}$ is a base for the topology on Aut $\mathcal{A}$ given by the standard metric. Now Aut $\mathcal{A}$ acts on P in a natural way, and with regard to this action we may discuss the homogeneity of P itself. We say that p and q in P are *compatible* if they have a common upper bound in P. We can now give a condition sufficient to rule out the third possibility above.

Theorem 5.1. *Suppose that* $\mathcal{A}$ *is a homogeneous structure such that for any* $p,q \in P$ *there is* $\theta \in \text{Aut}\,\mathcal{A}$ *such that* p *and the image of* q *under* θ *are compatible. Then any conjugacy class of* Aut $\mathcal{A}$ *which is comeagre on some non-empty open set is comeagre.*

Once one has formulated the condition given (which is just the usual notion of "homogeneity" in a set-theoretic context) the proof of the theorem is straightforward. It is likely that under the same or some strengthened hypotheses one will be able to prove that generic elements necessarily exist. Certainly there is a natural method for trying to construct a generic element σ as a union of a chain

$$\bigcup_{n\in\omega} p_n\,,\ p_n \in P \text{ all } n,$$

where we ensure that the nth element of A lies in

$$\text{dom } p_n \cap \text{range } p_n$$

and that for every $p \in P$ there is $q \in \text{Aut}\,\mathcal{A}$ taking p to an element of P below p_n for some n.

Our principal results at present consist in showing that certain groups have generic elements; moreover in each such case they may be constructed as just indicated.

Theorem 5.2. *The generic elements of* Symm Ω *(for* $|\Omega| = \aleph_0$*) are precisely those having cycle type* $\prod_{1\le k<\infty} k^{\infty}$.

Proof. Elements of this type certainly form a conjugacy class, so we just have to show that this conjugacy class is comeagre. The relevant dense open sets here are

(i) $D(x) = \{\sigma \in \text{Symm}\,\Omega : x \text{ is in a finite cycle of } \sigma\}$

and

(ii) $D(m,n) = \{\sigma \in \text{Symm}\,\Omega : \sigma \text{ has at least m cycles of length n}\}$.

Clearly a member of Symm Ω has cycle type as indicated if and only if it lies in

$$\bigcap\{D(x) : x \in \Omega\} \cap \bigcap\{D(m,n) : m,n \geq 1\}.$$

For Aut Γ_C it is not quite so easy to say *what* the generic elements should be, or to show that they *are* indeed generic.

Theorem 5.3. *The element* σ *of* Aut Γ_C *is generic if and only if*

(i) σ *has cycle type* $\prod_{1\leq k<\infty} k^{\infty}$,

(ii) *for any finite union* X *of cycles of* σ, $n \in \mathbf{N}$, *and maps*

$$\alpha: X \to C,\ \lambda: \{1,2,...,[n/2]\} \to C$$

such that $\sigma^n\alpha = \alpha$, *there is* $x \in \Gamma_C - X$ *lying in a cycle of* σ *of length* n *such that*

$$(\forall y \in X)\ F\{x,y\} = \alpha(y)\ \&\ (\forall i \in \{1,2,...,[n/2]\})\ F\{x,\sigma^i x\} = \lambda(i).$$

Here [n/2] is the integer part of n/2, and $\sigma^n\alpha$ is the map from X to C defined by

$$(\sigma^n\alpha)(y) = \alpha(\sigma^{-n}y)$$

all $y \in X$. The point of the rather obscure conditions of the theorem is that this is the most we can hope to require of the "witness" x. For example, if x lies in a cycle of length n and $(\forall y \in X)\ F\{x,y\} = \alpha(y)$ then

$$(\sigma^n\alpha)(y) = \alpha(\sigma^{-n}y) = F\{x,\sigma^{-n}y\} = F\{\sigma^n x,y\} = F\{x,y\} = \alpha(y),$$

so that $\sigma^n\alpha = \alpha$, and this has to be assumed.

The proof parallels that of Theorem 5.2. In this case however the details are not quite so trivial, and in particular rather a lot of effort is required to show that if $p \in P$ and $x \in \Gamma_C$ then p can be extended to q so that x is in a finite cycle of q. It is in fact the case that any $p \in P$ can be extended to $q \in P$ which is "tidy", in the sense that all its partial cycles are completed, or to put this another way, such that dom q = range q. Other concepts of "tidiness" may also be relevant for various purposes, in particular for studying least number of conjugates problems. See below for this.

The fact that the map λ is only defined on $\{1,2,...,[n/2]\}$ has to do with the fact that we are dealing with an undirected graph. This means that for any $i \in \{1,2,...,n-1\}$,

$$F\{x,\sigma^i x\} = F\{\sigma^n x,\sigma^i x\} = F\{\sigma^{n-i}x,x\} = F\{x,\sigma^{n-i}x\}.$$

For the corresponding countable universal C-coloured digraph one will expect instead λ to be defined on $\{1,2,\ldots,n-1\}$.

So much for some cases where there are lots of cycles of finite length. In A(Q) however, which is torsion-free, there are none, but we are still able to characterize the generic elements.

Theorem 5.4. *The member σ of* A(Q) *is generic if and only if the families of orbitals of σ of parities* +1,-1, *and* 0 *are each densely linearly ordered without endpoints and are each dense in the union of the others.*

This is intuitively plausible, on the following grounds. The information which some $p \in P$ can give us about $\sigma \in A(Q)$ is as follows:

If $px = x$ then p tells us that x is a fixed point of σ.

If $px > x$ then p tells us that $[x,px]$ is contained in an orbital of σ of parity +1.

If $px < x$ then p tells us that $[px,x]$ is contained in an orbital of σ of parity -1.

On the other hand p cannot tell us any *more* than this in the sense that any behaviour of σ not explicitly contradicted by the information provided in these three forms can actually occur for some σ extending p. This means that a generic element of A(Q) must satisfy the properties given in the theorem. Conversely it is easy to see by a back-and-forth argument that any two such elements must be conjugate.

We now move on to consideration of Aut Q, the group of homeomorphisms of Q to itself. Now this group is not closed as a subgroup of Symm Q (in fact it is dense) and is not a complete metric space under the usual metric. Indeed we may explicitly describe a failure of the Baire category theorem here by considering

$$D_n = \{\sigma \in \text{Aut } Q : \sigma(1/(n+1),1/n) \cap (n,n+1) \neq \emptyset\}.$$

Then each D_n is dense open but

$$\bigcap_{n\in\omega} D_n = \emptyset.$$

Since our previous definition of "generic" will not serve we instead return to our idea of using other metrics where appropriate. In this case the metric we use is given by

$d(\sigma,\tau) =$

$$\Sigma\{2^{-n} : \sigma q_n \neq \tau q_n \text{ or } \sigma^{-1}q_n \neq \tau^{-1}q_n\} + \Sigma\{2^{-n} : \sigma b_n \neq \tau b_n \text{ or } \sigma^{-1}b_n \neq \tau^{-1}b_n\}$$

where $\{q_n : n \in \omega\}$ enumerates **Q** and $\{b_n : n \in \omega\}$ is a family of clopen subsets of **Q** forming a base for the (usual) topology and such that $\mathbf{B} = \{b_n : n \in \omega\}$ is a Boolean algebra. This metric was suggested by consideration of the proof given in [24] of the uniqueness of the rational topology (subject to certain conditions). Here a base for the topology derived from the metric is given by sets of the form

$$[p,q] = \{\sigma \in \text{Aut } \mathbf{Q} : \sigma \text{ extends p and } \sigma \text{ induces q on its domain}\}$$

where p is a finite partial bijection of **Q** with dom p = range p, and q is an automorphism of a finite Boolean subalgebra of **B**, such that

$$x \in \text{dom p} \ \& \ b \in \text{dom q} \ \& \ x \in b \rightarrow px \in qb.$$

It is not hard to see that d is a complete metric on Aut **Q**, so that the definition of "generic" makes sense. Moreover the sets [p,q] are all open, and they correspond to the natural method for approximating members of Aut **Q**, - p giving finitely many of the values of σ and q restricting what future values can be.

Theorem 5.5. *With respect to this metric,* Aut **Q** *has generic elements, all of them having cycle type* $\prod_{1 \leq k < \infty} k^{\infty}$

A more exact description of what these elements are will be given in [27] and [28].

6. Reducing the number of conjugates

We conclude by making a few rather speculative remarks about how one might hope to obtain optimal "number of conjugates" results for various simple (or nearly simple) permutation groups. This might appear to be of purely technical interest, and it is certainly true that it is more important to know that there *is* a bound rather than being sure what the best one is; all the same we incline to the view that it is not an entirely idle question. Possibly this minimal number may tell us something important about the structure - and that is certainly our hope. In a few cases we know the exact best number, but generally not. For example in Alt Ω for infinite Ω there is no bound at all on the number of conjugates required. In A(Q), for σ,τ with $\tau \in \langle\sigma\rangle^G$, the best one can say in general is that at most 4 conjugates (of σ or σ^{-1}) are required, though for σ or τ in the various normal subgroups, 2 or 3 may suffice.

The interesting test case once again seems to be Aut Γ_2 (Γ_2 the random graph). Here Rubin [18] has shown that 4 conjugates suffice but 2 do not. One can see the latter by an argument involving just cycle types as follows. According to [23] Aut Γ_2 has elements σ and τ of cycle type 2^∞ and ∞^1 respectively and it was shown in [26] that for such elements, τ cannot be expressed as the product of any two conjugates of σ.

In an effort to discover the optimum number (3 or 4) one is led to try to find the correct choice of Σ, the family of "intermediate" elements. One would like to show that

(i) for any $\sigma_1,\sigma_2 \neq 1$, there are conjugates τ_1,τ_2 of σ_1,σ_2 such that $\tau_2\tau_1 \in \Sigma$,

(ii) any two members of Ω are conjugate.

From this one could at once deduce that three conjugates suffice. To date three possible choices of Σ have been proposed, for none of which do we obtain both (i) and (ii). For the first in [23] we took σ to lie in Σ provided it has cycle type ∞^∞ and for any finite disjoint U,V $\subseteq \Gamma_2$ there is $x \in \Gamma_2$ such that the σ-cycle of x is disjoint from $U \cup V$ and x is joined to every member of U and to none of V. Now property (i) is readily established, but in place of property (ii) we can only show that if $\sigma_1,\sigma_2,\sigma_3 \in \Sigma$ there are conjugates τ_i of σ_i such that $\tau_3\tau_2\tau_1 = 1$. This establishes the sufficiency of 5 conjugates instead of the desired 3. Moreover it is certainly not the case that all members of Σ are conjugate.

Rubin in [18] greatly strengthened the requirement for σ to lie in Σ, calling this notion *uniform*. The outcome was that any two uniform elements are now conjugate, so that (ii) holds, but it is not so easy to obtain uniform elements as products of conjugates of arbitrary non-identity elements. Thus 3 conjugates instead of 2 are required for (i). The result here still therefore refers to 5 conjugates (this time as 3+3-1 rather than as 2+2+2-1) but it has the merit of applying in much greater generality. As a third possibility Rubin also considered a class of elements called *regular*, somewhere intermediate between the other two. These are not now all conjugate, but (i) holds in its original form, and a modification of (ii) allows one to reduce the 5 conjugates to 4.

A tempting idea is to try to use generic elements of Aut Γ_C as the members of Σ, since (by definition) they are all conjugate, but it is then easy to see that (i) cannot hold since all generic elements have fixed points.

I should like to suggest that the correct way to tackle this problem is to omit the intermediate step and to try to prove outright that for any non-identity

$\sigma_1,\sigma_2,\sigma_3,\sigma_4 \in$ Aut Γ_2

there are conjugates $\tau_1,\tau_2,\tau_3,\tau_4$ such that $\tau_4\tau_3\tau_2\tau_1 = 1$, establishing the sufficiency of 3 conjugates. (It is best to hold all 4 permutations on the left hand side for reasons of symmetry). One will use suitable quadruples (p_1,p_2,p_3,p_4) of finite partial automorphisms approximating conjugacies θ_i such that $\tau_i = \theta_i\sigma_i\theta_i^{-1}$. A quadruple (x_1,x_2,x_3,x_4) is said to be *linked* if

$$x_i, \sigma_i x_i \in \text{dom } p_i$$

for each i and

$$p_i\sigma_i x_i = p_{i+1}x_{i+1}$$

for each i (where the suffices are taken mod 4). The idea is that saying (x_1,x_2,x_3,x_4) is linked is equivalent to saying that

$$\tau_4\tau_3\tau_2\tau_1(p_1x_1) = p_1x_1$$

whenever $(\theta_1,\theta_2,\theta_3,\theta_4)$ is compatible with (p_1,p_2,p_3,p_4). So the crucial point is to ensure that for every approximation we allow and every specified member of Γ_2 there is an allowed extension in which that element occurs in a linked quadruple (at a particular entry). The problem is to make the right choice of approximation to ensure that this can always be done. The trouble is that if one extends without due care so as to include the element in a linked quadruple then this can destroy the chances for other elements. These difficulties are particularly acute when one or more of the σ_i has cycles of length 1 or 2. What is needed is a good way to handle the priorities for elements' being included in the domains of the various p_i without prejudicing the chances for other elements in the future.

In conclusion the moral arising from considerations of this sort, and of most of the remarks we have made about infinite simple groups is this. The structures we are interested in are particularly homogeneous and can normally be viewed as having been constructed by means of a sequence of finite approximations of some sort. It is then very often possible and desirable to view their automorphism groups in a similar light - via an appropriate approximation structure, though here the computations may have to be a good deal more delicate. The way in which the combinatorics work out may well provide information about the group - what its normal subgroups are, minimum number of conjugates results, how much of the structure can be recovered from the group - but in exploring this one may also come to achieve a much more thorough understanding of the structure itself.

References

1. R D Anderson, The algebraic simplicity of certain groups of homeomorphisms, *Amer. J. Math.* **80** (1958), 955-963.

2. R N Ball & M Droste, Normal subgroups of doubly transitive automorphism groups of chains, *Trans. Amer. Math. Soc.* **290** (1985), 647-664.

3. P Bruyns, *Aspects of the group of homeomorphisms of the rational numbers* (Doctoral thesis, Oxford 1986).

4. M Droste, The normal subgroup lattice of 2-transitive automorphism groups of linearly ordered sets, *Order* **2** (1985), 291-319.

5. M Droste & S Shelah, A construction of all normal subgroup lattices of 2-transitive automorphism groups of linearly ordered sets, *Israel J. Math.* **51** (1985), 223-261.

6. M Droste & J K Truss, Subgroups of small index in ordered permutation groups, *Quart. J. Math.*, to appear.

7. R Fraïssé, Sur l'extension aux rélations de quelques propriétés des ordres, *Ann. Sci. Ecole Norm. Sup.***71** (1954), 361-388.

8. A M W Glass, *Ordered permutation groups* (London Math. Soc. Lecture Notes **55** Cambridge Univ. Press, London 1981).

9. Y Gurevich & W C Holland, Recognizing the real line, *Trans. Amer. Math. Soc.* **265** (1981), 527-534.

10. C W Henson, Countable homogeneous structures and $\aleph_0$-categorical theories, *J. Symbolic Logic* **37** (1972), 494-500.

11. G Higman, On infinite simple groups, *Publ. Math. Debrecen* **3** (1954), 221-226.

12. M Jambu-Giraudet, *Théorie des modèles de groupes d'automorphismes d'ensembles totalement ordonnés* (Thèse 3eme Cycle, Université Paris VII, Paris, 1979).

13. M Jambu-Giraudet, Bi-interpretable groups and lattices, *Trans. Amer. Math. Soc.* **278** (1983), 253-269.

14. M Jambu-Giraudet, Quelques remarques sur l'équivalence élémentaire entre groupes ou treillis d'automorphismes de chaines 2-homogènes, *Discrete Math.* **53** (1985), 117-124.

15. A Lachlan & R Woodrow, Countable ultrahomogeneous graphs, *Trans. Amer. Math. Soc.* **262** (1980), 51-94.

16. A Mekler, Groups embeddable in the autohomeomorphisms of **Q**, *J. London Math. Soc.* **33** (1986), 49-58.

17. P M Neumann, Automorphisms of the rational world, *J. London Math. Soc.* **32** (1985), 439-448.

18. M Rubin, Untitled manuscript, 1987.

19. M Rubin, On the reconstruction of $\aleph_0$-categorical structures from their automorphism groups, preprint 1988.

20. M Rubin, On the reconstruction of Boolean algebras from their automorphism groups, *Handbook of Boolean Algebras* (ed. Monk, Elsevier 1989).

21. M Rubin, On the reconstruction of topological spaces from their groups of homeomorphisms, *Trans. Amer. Math. Soc.* **312** (1989), 487-538.

22. S Shelah, Can you take Solovay's inaccessible away? *Israel J. Math.* **48** (1984), 1-47.

23. J K Truss, The group of the countable universal graph, *Math. Proc. Cambridge Philos.Soc.* **98** (1985), 213-245.

24. J K Truss, Embeddings of infinite permutation groups, *Proceedings of Groups-St Andrews 1985* (Cambridge University Press 1986), 335-351.

25. J K Truss, The group of almost automorphisms of the countable universal graph, *Math. Proc. Cambridge Philos. Soc.* **105** (1989), 223-236.

26. J K Truss, Infinite permutation groups;products of conjugacy classes, *J. Algebra* **120** (1989), 454-493.

27. J K Truss, Conjugate homeomorphisms of the rationals, to appear.

28. J K Truss, Generic automorphisms of homogeneous structures, to appear.

POLYNOMIAL 2-COCYCLES

L R VERMANI

Kurukshetra University, Kurukshetra 132119, India

1. Introduction

Let G be a group, ZG the integral group ring of G and $\Delta(G)$ its augmentation ideal. Recall that when A is a trivial G-module, a 2-cocycle, $f : G \times G \to A$ is a polynomial 2-cocycle of degree $\leq k$ if the extension of f by linearity to $ZG \times ZG$ vanishes on $\Delta^{k+1}(G) \times ZG$ (or equivalently on $ZG \times \Delta^{k+1}(G)$). Let $P_kH^2(G,A)$ denote the subgroup of $H^2(G,A)$ consisting of all elements ξ which possess a polynomial 2-cocycle of degree $\leq k$ as representative. The study of the subgroups $P_kH^2(G,T)$, where T is the additive group of rationals mod 1 regarded as a trivial G-module, is closely related to the study of dimension subgroups - polynomial 2-cocycles arose in Passi's study [10] of the dimension subgroup problem.

If G is a finite p-group, there exists [10] an integer n such that $P_nH^2(G,T) = H^2(G,T)$. However, it is not known whether such an n exists for every nilpotent group G. Some positive results in this direction have been obtained by Passi-Vermani [19], Passi-Sucheta [16] and Passi-Sucheta-Tahara [17]. In the present paper we review some of the work on polynomial 2-cocycles concerning this problem.

2. Elements of finite degree

First we recall a useful observation which follows from a technical result of Passi ([12, Theorem 2.1]).

Let

$$1 \to A \to \pi \to G \to 1 \tag{2.1}$$

be a central extension of an abelian group A by a group G and ξ be the element of $H^2(G,A)$ which corresponds to the central extension (2.1).

Proposition 2.2. *If the identity map* : $A \to A$ *can be extended to a map* $\pi \to A$ *the linear extension of which to* $Z\pi$ *vanishes on*

$$\Delta^{n+2}(\pi) + \Delta(\pi)\,\Delta(A),$$

then $\xi \in P_nH^2(G,A)$.

Using this result Passi-Vermani [19] obtain a criterion for identifying central extensions of a divisible abelian group A by G which correspond to elements of $P_nH^2(G,A)$. Since the original paper does not contain a proof, it is worthwhile to give one here.

Proposition 2.3. *If the group* A *in* (2.1) *is divisible abelian, then* $\xi \in P_nH^2(G,A)$ *if and only if*

$$A \cap (1 + \Delta^{n+2}(\pi) + \Delta(\pi)\Delta(A)) = 1.$$

Proof. Let $\xi \in P_nH^2(G,A)$ and let $\{w(g)\}_{g\in G}$, $w(1) = 1$, be a transversal for G in π such that the corresponding 2-cocycle $W : G \times G \to A$ is a polynomial 2-cocycle of degree $\leq n$. Every element of π can be uniquely written as $w(g)a$, $g \in G$, $a \in A$. Define a map $\psi : \pi \to A$ by setting $\psi(w(g)a) = a$, $g\in G$, $a \in A$. Then it turns out that the extension of Ψ to $Z\pi$ by linearity vanishes on

$$\Delta^{n+2}(\pi) + \Delta(\pi)\Delta(A).$$

It follows that $A \cap (1 + \Delta^{n+2}(\pi) + \Delta(\pi)\Delta(A)) = 1$.

Conversely, suppose that $A \cap (1 + \Delta^{n+2}(\pi) + \Delta(\pi)\Delta(A)) = 1$. Then the homomorphism

$$\alpha : A \to Z\pi/(\Delta^{n+2}(\pi) + \Delta(\pi)\Delta(A))$$

given by $\alpha(a) = a - 1 + \Delta^{n+2}(\pi) + \Delta(\pi)\Delta(A)$, $a \in A$, is a monomorphism. The group A being divisible, α splits. Thus the identity map: $A \to A$ has an extension $\psi : \pi \to A$ the extension of which to $Z\pi$ by linearity vanishes on $\Delta^{n+2}(\pi) + \Delta(\pi)\Delta(A)$. The result then follows from Proposition 2.2.

Observe that $\Delta(\pi)\Delta(A) \subseteq \Delta^n(\pi)$ for all $n > 1$ provided π is a torsion group and A is divisible abelian. As such we have:

Corollary 2.4. *If* G *is a torsion group and* A *a torsion divisible abelian group, then* $\xi \in P_nH^2(G,A)$ *if and only if* $A \cap D_{n+2}(\pi) = 1$.

Another useful criterion for an element corresponding to the central extension (2.1) to be of degree $\leq n$ is given by:

Proposition 2.5. *The element* $\xi \in P_nH^2(G,A)$ *if and only if*

$$\Delta^{n+2}(\pi) \cap Z\pi\, \Delta(A) \leq \Delta(\pi)\Delta(A).$$

Proof. Observe that the homomorphism

$$\alpha : A \to Z\pi/(\Delta^{n+2}(\pi) + \Delta(\pi)\Delta(A))$$

given by $a \to a - 1 + \Delta^{n+2}(\pi) + \Delta(\pi)\Delta(A)$, $a \in A$, is a monomorphism if and only $A \cap (1 + \Delta^{n+2}(\pi) + \Delta(\pi)\Delta(A)) = 1$. On the other hand $\Delta^{n+2}(\pi) \cap Z\pi\Delta(A) \leq \Delta(\pi)\Delta(A)$ holds if and only if the natural homomorphism.

$$\beta : Z\pi\Delta(A)/\Delta(\pi)\Delta(A) \to Z\pi/(\Delta^{n+2}(\pi) + \Delta(\pi)\Delta(A))$$

is one-one. Since $\pi \cap (1 + \Delta(\pi)\Delta^m(A)) = D_{m+1}(A)$ for all $m \geq 1$ [20], the map $a \to a - 1 + \Delta(\pi)\Delta(A)$, $a \in A$, is an isomorphism $\gamma : A \to Z\pi\Delta(A)/\Delta(\pi)\Delta(A)$. But $\alpha = \beta\gamma$ and the result follows from Proposition 2.3.

3.

Let

$$1 \to N \to G \to K \to 1 \tag{3.1}$$

be an exact sequence of groups. Denote by $E_n(N,G)$ the subgroup of G given by

$$E_n(N,G)/[N,G] = G/[N,G] \cap (1 + \Delta^n(G/[N,G]) + \Delta(G/[N,G])\Delta(N/[N,G])).$$

Then $\{E_n(N,G)\}$ is a decreasing sequence of normal subgroups of G.

The terms of low degree of the Hochschild-Lyndon-Serre spectral sequence corresponding to the exact sequence (3.1) give a 5-term exact sequence. Using this 5-term sequence, Passi-Sharma-Vermani [14] obtain:

Theorem 3.2. *For every* $n \geq 0$ *and divisible abelian group* A *regarded as a trivial* G*-module, there exists a natural exact sequence* $0 \to \mathrm{Hom}(K/K', A) \to$

$\mathrm{Hom}(G/G',A) \xrightarrow{\alpha} \mathrm{Hom}(N/N \cap E_{n+2}(N,G),A) \xrightarrow{t} P_nH^2(K,A) \to P_nH^2(G,A)$ *where* $P_nH^2(K,A) \to P_nH^2(G,A)$ *is the restriction of the inflation homomorphism* $H^2(K,A) \to H^2(G,A)$ *to the subgroup* $P_nH^2(K,A)$.

As an application of this theorem we obtain:

Theorem 3.3 ([14]). *Let* $1 \to R \to F \to G \to 1$ *be a free presentation of a group* G, n *a positive integer and* A *a divisible abelian group regarded as a trivial* G*-module. Then*

$$P_nH^2(G,A) \cong \mathrm{Hom}(R \cap F/R \cap E_{n+2}(R,F), A).$$

Writing $S[R,F]/[R,F] = \overline{S}$ for a subgroup S of F one obtains alternative proofs of the following results of Passi-Stammbach [15] and of Passi [10].

Corollary 3.4 ([15]). *For any positive integer* n, $P_nH^2(G,T) = H^2(G,T)$ *if and only if* $\bar{R} \cap (1 + \Delta^{n+2}(\bar{F}) + \Delta(\bar{F})\Delta(\bar{R})) = 1$.

Corollary 3.5. *If* $G = F/\gamma_{n+1}(F)$ *is a free nilpotent group of class* n, *then* $P_mH^2(G,T) = 0$ *for* $m < n$ *and* $P_nH^2(G,T) = H^2(G,T)$.

Using Corollary 3.4 and a result of Sandling [13], we get an alternative proof of Passi's result that

$$P_1H^2(G,T) = H^2(G,T) \text{ for every abelian group } G. \tag{3.6}$$

We may point out that the above results have been obtained in [14] for the varietal multiplicator.

Recall that an ideal I of a right Noetherian ring R has weak Artin-Rees property if for each finitely generated right R-module M and submodule U of M there exists a positive integer n with $MI^n \cap U \leq UI$. Every polycentral ideal of a right Noetherian ring has weak Artin-Rees property [9] and it is quite easily proved that $\Delta(G)$ is a polycentral ideal of **Z**G if G is a finitely generated nilpotent group. Using this observation Passi-Vermani [19] obtained:

Theorem 3.7. *If* G *is a finitely generated nilpotent group, then there exists an integer* n *such that* $P_nH^2(G,T) = H^2(G,T)$.

This meant an improvement on the earlier result of Passi [10] obtained for finite p-groups.

Using a lemma of Hartley ([4, Lemma 4.1]) and Proposition 2.5, Passi-Vermani [19] proved:

Theorem 3.8. *If* G *is a torsion free nilpotent group of class* c *then*

$$P_nH^2(G,T) = H^2(G,T), \text{ where } n = c^2 + 4c + 1.$$

However, first improving the lemma of Hartley, Passi-Sucheta [16] prove the above result with $n = 3c+1$. Using the fact that divisible nilpotent groups have the dimension property [16] they also proved:

Theorem 3.9. *If* G *is a divisible nilpotent group of class* c, *then*

$$P_cH^2(G,T) = H^2(G,T).$$

We can prove independently of Röhl [21] the existence of an n such that $P_nH^2(G,T) = H^2(G,T)$ for a divisible nilpotent group G (much weaker than Theorem 3.9). Such a proof uses the six term exact sequence of Iwahori-Matsumoto [8]. Also replacing T by Q, the additive group of rationals regarded as a trivial G-module, Passi-Sucheta-Tahara [17] prove that

$$P_nH^2(G,Q) = H^2(G,Q)$$

if G is a torsion free nilpotent group of class n and then use it to prove that $P_nH^2(G,Q) = H^2(G,Q)$ when G is a divisible nilpotent group of class n.

N D Gupta [3] has proved that finitely generated metabelian p-groups, $p \neq 2$, have the dimension property. It will be interesting to find its effect on $P_nH^2(-,T)$. Explicitly:

Problem 3.10. If G is a finitely generated, nilpotent, metabelian p-group, $p \neq 2$, is $P_cH^2(G,T) = H^2(G,T)$, where c is the nilpotency class of G?

4.

Theorem 4.1 ([18],[23]). *For every group* G *and integer* $n \geq 1$, *there is an exact sequence*

$$0 \to \mathrm{Hom}(\Delta^2(G)/\Delta^{n+2}(G),T) \to \mathrm{Hom}(\Delta(G)/\Delta^{n+1}(G) \otimes_G \Delta(G)/\Delta^{n+1}(G),T)$$

$$\to P_nH^2(G,T) \to 0.$$

An application of this result gives $P_1H^2(G,T) = H^2(G,T)$ if G is an extra-special p-group, p an odd prime [23] giving infinitely many nilpotent groups G of class 2 with every element of $H^2(G,T)$ of degree ≤ 1. Such a situation can also arise for non-nilpotent groups as well.

Example. Let $G = G_1 * G_2$ be the free product of groups G_1, G_2 and A be a trivial G-module. Then

$$H^2(G,A) \cong H^2(G_1,A) \oplus H^2(G_2,A).$$

An isomorphism : $H^2(G,A) \to H^2(G_1,A) \oplus H^2(G_2,A)$ is induced by the restriction homomorphisms : $H^2(G,A) \to H^2(G_i,A)$, $i = 1,2$. Restricting this isomorphism to $P_nH^2(G,A)$ we get an isomorphism [23]

$$P_nH^2(G,A) \to P_nH^2(G_1,A) \oplus P_nH^2(G_2,A). \quad (4.2)$$

Now taking G_1, G_2 to be non-cyclic abelian groups, A = T and using (3.6) we find that $P_1H^2(G,T) = H^2(G,T)$.

The similarity observed in the behaviour of $P_nH^2(-,A)$ and $H^2(-,A)$ with regard to the free product is also seen for direct sums. We give this in the following unpublished result proved jointly with Passi.

Theorem 4.3. *Let* $G = H \oplus K$ *and* A *be a trivial* G-*module. Then there exists an isomorphism*

$$H^2(G,A) \to H^2(H,A) \oplus H^2(K,A) \oplus \mathrm{Hom}(H/\gamma_2(H) \otimes K/\gamma_2(K),A)$$

the restriction of which to $P_nH^2(G,A)$ *induces an isomorphism*

$$P_nH^2(G,A) \cong P_nH^2(H,A) \oplus P_nH^2(K,A) \oplus \mathrm{Hom}(H/\gamma_2(H) \otimes K/\gamma_2(K),A).$$

Proof. If f is a 2-cocycle, we denote its cohomology class by [f]. We also identify a homomorphism $g : H/\gamma_2(H) \otimes K/\gamma_2(K) \to A$ with the corresponding bilinear map : $H \times K \to A$.

Let $\alpha:H^2(G,A) \to H^2(H,A)$, $\beta:H^2(G,A) \to H^2(K,A)$ denote the restriction homomorphisms. Let $\xi \in H^2(G,A)$ and $f : G \times G \to A$ be a representative 2-cocycle for ξ. Define $g : H \times K \to A$ by

$$g(x,y) = f(y,x) - f(x,y), \ x \in H, \ y \in K.$$

The map g is independent of the choice of the 2-cocycle f representing ξ and is bilinear. Define a map

$$\theta : H^2(G,A) \to H^2(H,A) \oplus H^2(K,A) \oplus \mathrm{Hom}(H/\gamma_2(H) \otimes K/\gamma_2(K),A)$$

by

$$\theta(\xi) = (\alpha(\xi), \beta(\xi), g), \qquad \xi \in H^2(G,A).$$

The map θ is a homomorphism and maps $P_nH^2(G,A)$ into

$$P_nH^2(H,A) \oplus P_nH^2(K,A) \oplus \mathrm{Hom}(H/\gamma_2(H) \otimes K/\gamma_2(K),A).$$

Next, let $\xi_1 \in H^2(H,A)$, $\xi_2 \in H^2(K,A)$ and $g \in \mathrm{Hom}(H/\gamma_2H) \otimes K/\gamma_2(K),A)$.

Let $\xi_i = [f_i]$, $i = 1,2$. Define $f : G \times G \to A$ by

$$f(x_1y_1, x_2y_2) = f_1(x_1,x_2) + f_2(y_1,y_2) + g(x_2,y_1), \ x_i \in H, \ y_i \in K, \ i = 1,2.$$

The map f is a 2-cocycle and [f] is independent of the choice of the 2-cocycle representatives f_1,f_2 of ξ_1, ξ_2 respectively. Moreover, if f_1, f_2 are both polynomial 2-cocycles of degree $\leq n$, then f is also easily seen to be a polynomial 2-cocycle of degree $\leq n$. We thus have a map

$$\phi : H^2(H,A) \oplus H^2(K,A) \oplus \mathrm{Hom}(H/\gamma_2(H) \otimes K/\gamma_2(K),A) \to H^2(G,A)$$

which is a homomorphism and which maps

$$P_nH^2(H,A) \oplus P_nH^2(K,A) \oplus \mathrm{Hom}(H/\gamma_2(H) \otimes K/\gamma_2(K),A)$$

into $P_nH^2(G,A)$. It is easy to check that θ and ϕ are inverses of each other.

The decomposition for $H^2(H \oplus K, F^*)$ given in the above theorem is available for finite groups in Huppert ([6, p.650]), where F^* is the multiplicative group of an algebraically closed field F of characteristic zero. For divisible abelian A, the decomposition of $H^2(H \oplus K,A)$ is a special case of the Künneth theorem ([5, p.222-3]).

5.

Passi [10] proved that if G is a finite p-group of class 2, $p \neq 2$, then $P_2H^2(G,T) = H^2(G,T)$. For a 2-group G of class 2 he proved [10] that $P_2H^2(G,T) = H^2(G,T)$ if $G/\gamma_2(G)$ is a direct sum of two cyclic groups or at most one direct summand of $G/\gamma_2(G)$ is of order ≥ 4. Working on the same lines Manju [2] proved Passi's result for G with $\gamma_2(G)$ cyclic or $G/\gamma_2(G)$ a direct sum of three cyclic groups at least two of which have the same order. Recently Passi-Sucheta-Tahara [17] removed the restriction of two direct summands of $G/\gamma_2(G)$ being of equal order from the result of Manju. However, the approach in [17] is quite different from the one adopted in [2] and [10]. In [17] the result is obtained by first proving:

Theorem 5.1. *If* G *is a finite 2-group with an* N*-series* $G = H_1 \geq H_2 \geq \ldots$ *such that* H_1/H_2 *is a direct sum of three cyclic groups of orders* $d(i) = 2^{\alpha_i}$, $1 \leq i \leq 3$, $\alpha_1 \leq \alpha_2 \leq \alpha_3$ *and the exponent of* H_2/H_3 *divides* d(2), *then* $G \cap (1 + \Lambda_4) = H_4$ *where* Λ_4 *denotes the 4th term in the canonical filtration of* $\Delta(G)$ *induced by the given* N*-series.*

Another interesting result that Passi-Sucheta-Tahara [17] prove is:

Theorem 5.2. *Let* G *be a nilpotent group of class* n *such that* H_2G (= *the* 2nd *integral homology group of* G) *is torsion free. Then* $P_nH^2(G,T) = H^2(G,T)$.

In view of this it will be interesting to investigate conditions under which H_2G, G nilpotent, is torsion free.

As a consequence of a result of Blackburn and Evens [1] we find that:

Proposition 5.3. *If* G *is a non-abelian group with a normal subgroup* n *such that both* N *and* G/N *are infinite cyclic, then* $H_2G = 0$.

Now let G be a group with a normal subgroup $N = \langle a,b \mid b^{-1}ab = a^r, r \neq 0,1 \rangle$ and $G/N = \langle gN \rangle$ an infinite cyclic group. Again using [1] we obtain:

Theorem 5.4. (i) *If* $G = N \oplus \langle g \rangle$ *and* $r \neq 2$, *then* H_2G *is a direct sum of a finite cyclic group and an infinite cyclic group.*

(ii) *If* $g = N \oplus \langle g \rangle$ *and* $r = 2$, *then* H_2G *is infinite cyclic.*

(iii) *If* $g^{-1}ag \neq a$ *but* $g^{-1}bg = b$, *then* H_2G *is infinite cyclic.*

(iv) *If* $g^{-1}ag = a$ *but* $g^{-1}bg \neq b$, *then* H_2G *is finite cyclic.*

(v) *If* $r = 2$ *and* $g^{-1}bg \neq b$, *then* $H_2G = 0$.

However, the subgroup N of G is not nilpotent and so the group G is not nilpotent.

We next mention just two results to indicate the relation between the study of $P_nH^2(G,T)$ and dimension subgroups. A true picture of this relationship can be had only by looking at the papers [10], [13] and [19]. It follows from Sjogren's work that for any p-group G, $D_n(G) = \gamma_n(G)$ for $n \leq p+1$. Using this result one obtains [19]:

Theorem5.5. *If* G *is a nilpotent* p*-group of class* $c < p$, *then* $P_cH^2(G,T) = H^2(G,T)$.

Then using this result we obtain [19]:

Corollary 5.6. *If* G *is a* p*-group, then*

$$G \cap (1 + \Delta^n(G) + \Delta(G)\Delta(\xi(G))) = \gamma_n(G), \text{ for } 1 \leq n \leq p+1.$$

Theorem 5.7 ([19]). *The following statements are equivalent:*

(i) *There exist constants* $c_1, c_2, \ldots, c_n, \ldots$ *such that for every group* G, $D_n(G)^{c_n} \leq \gamma_n(G)$ *for all* $n \geq 1$.

(ii) *There exist constants* $d_1, d_2, \ldots, d_n, \ldots$ *such that for every nilpotent group* G *of class* $\leq n$,

$$d_nH^2(G,T) \leq P_nH^2(G,T).$$

6.

The study of polynomial 2-cocycles is also useful in certain situations not concerning the dimension subgroups. For example Passi-Stammbach [15] have given a cohomological characterization of parafree groups using $P_nH^2(G,T)$.

Paul Igodt [7] proves:

Proposition 6.1. *If* G *is a free abelian group of finite rank and* A *is a free abelian group of finite rank regarded as a trivial* G*-module, then* $P_1H^2(G,A) = H^2(G,A)$.

He then uses it to prove that every torsion free nilpotent group of class 2 and rank k+n can in a constructive manner be embedded as a group of unipotent matrices in Gl(k+n+1,Z).

Passi (unpublished) has obtained the following generalization of Proposition 6.1.

Theorem 6.2. *Let* G *be a finitely generated nilpotent group of class* n *in which the quotients of the lower central series are torsion free and* A *be a free abelian group regarded as a trivial* G*-module. Then* $P_nH^2(G,A) = H^2(G,A)$.

Proof. Let $E : 1 \to A \to M \to G \to 1$ be a central extension of A by G. Now $A \cap (1 + \Delta^{n+2}(M) + \Delta(M)\Delta(A)) \subseteq M(1 + \Delta_Q^{n+2}(M) + \Delta_Q(\zeta(M))\Delta_Q(M)) = \sqrt{\gamma_{n+2}(M)} = \sqrt{1}$ (by [17]). Thus $A \cap (1 + \Delta^{n+2}(M) + \Delta(M)\Delta(A))$ is a torsion subgroup of A and hence is 1. Thus the map

$$\alpha: A \to ZM/(\Delta^{n+2}(M) + \Delta(M)\Delta(A))$$

given by

$$a \to a - 1 + \Delta^{n+2}(M) + \Delta(M)\Delta(A),\ a \in A,$$

is a monomorphism. It follows from Theorem 3.2 of Hartley [4] that the cokernel $ZG/\Delta^{n+2}(G)$ of this homomorphism is torsion free. Therefore the exact sequence

$$1 \to A \xrightarrow{\alpha} ZM/(\Delta^{n+2}(M) + \Delta(M)\Delta(A)) \to ZG/\Delta^{n+2}(G) \to 1$$

splits and the identity map $i : A \to A$ can be extended to a map : $M \to A$ the extension of which by linearity to ZM vanishes on $\Delta^{n+2}(M) + \Delta(M)\Delta(A)$. It follows from Proposition 2.2 that E represents an element of degree $\leq n$.

References

1. N Blackburn & L Evens, Schur multipliers of p-groups, *J. Reine Angew. Math.* **309** (1979), 100-113.
2. M Goyal, *The augmentation ideal of an integral group ring* (Ph.D. Thesis, Kurukshetra University, Kurukshetra, India, 1981).
3. N D Gupta, The dimension subgroup conjecture, *Bull. Amer. Math. Soc.*, to appear.
4. B Hartley, Powers of the augmentation ideal in group rings of infinite nilpotent groups, *J. London Math. Soc.* **25**(2) (1982), 43-61.
5. P J Hilton & U Stammbach, *A Course in Homological Algebra* (Springer-Verlag, Berlin-Heidelberg-New York, 1970).
6. B Huppert, *Endliche Gruppen I* (Springer-Verlag, Berlin-Heidelberg-New York, 1967).
7. P Igodt, Torsion free 2-nilpotent groups and polynomial 2-cocycles, *Nederl. Akad. Wetensch. Proc. Ser. A* **87**(4) (1984), 415-420.
8. N Iwahori & H Matsumoto, Several remarks on projective representations of finite groups, *J. Fac. Sci. Univ. Tokyo* **10** (1963/64), 129-146.

9. D S Passman, *Algebraic Structures of Group Rings* (John Wiley and Sons, 1977).

10. I B S Passi, Dimension subgroups, *J. Algebra* **9** (1968), 152-182.

11. I B S Passi, Induced central extensions, *J. Algebra* **16** (1970), 27-39.

12. I B S Passi, Polynomial maps on groups II, *Math. Z.* **135** (1974), 137-141.

13. I B S Passi, *Group rings and their augmentation ideals* (Lecture Notes in Mathematics **715**, Springer-Verlag, Berlin-Heidelberg-New York, 1979).

14. I B S Passi, S Sharma & L R Vermani, A filtration of varietal multiplicator, *Proc. Nat. Acad. Sci. India* (P L Bhatnagar Commemorative Volume) (1979), 377-388.

15. I B S Passi & U Stammbach, A filtration of Schur multiplicator, *Math. Z.* **135** (1974), 143-148.

16. I B S Passi & Sucheta, Dimension subgroups and Schur multiplicator II, *Topology Appl.* **25** (1987), 121-124.

17. I B S Passi, Sucheta & K Tahara, Dimension subgroups and Schur multiplicator III, *Japan J. Math.* **19** (1987), 371-379.

18. I B S Passi & L R Vermani, The inflation homomorphism, *J. London Math. Soc.* (2) **6** (1972), 129-136.

19. I B S Passi & L R Vermani, Dimension subgroups and Schur multiplicator, *J. Pure Appl. Algebra* **30** (1983), 61-67.

20. Ram Karan & L R Vermani, A note on augmentation quotients, *Bull. London Math. Soc.* **18** (1986), 5-6.

21. F Röhl, On group ring quotients of divisible groups, *Arch. Math.* **40** (1983), 297-303.

22. U Stammbach, *Homology in Group Theory* (Lecture Notes in Mathematics **359**, Springer-Verlag, Berlin-Heidelberg-New York, 1973).

23. L R Vermani, The Schur's multiplicator-II, *Math. Student* **55** (1987), 151-158.

Printed in the United States
By Bookmasters